CSCPRC REPORT NO. 3

Pure and Applied Mathematics in the People's Republic of China

A Trip Report of the American Pure and Applied Mathematics Delegation

Edited by ANNE FITZGERALD and SAUNDERS MAC LANE

Submitted to the Committee on Scholarly Communication with the People's Republic of China

NATIONAL ACADEMY OF SCIENCES
Washington, D.C. 1977

The exchange visit of the Pure and Applied Mathematics Delegation to the People's Republic of China was supported by a grant from the National Science Foundation. This visit was part of the exchange program operated by the Committee on Scholarly Communication with the People's Republic of China, founded jointly in 1966 by the American Council of Learned Societies, the National Academy of Sciences, and the Social Science Research Council. Sources of funding for the Committee include the National Science Foundation, the Department of State, and the Ford Foundation.

The Committee represents American scholars in the natural, medical, and social sciences, as well as the humanities. It advises individuals and institutions on means of communicating with their Chinese colleagues, on China's international scholarly activities, and on the state of China's scientific and scholarly pursuits. Members of the Committee are scholars from a broad range of fields, including China studies.

Administrative offices of the Committee are located at the National Academy of Sciences, Washington, D.C.

Library of Congress Catalog Card Number 77-79329

International Standard Book Number 0-309-02609-1

Available from
Printing and Publishing Office
National Academy of Sciences
2101 Constitution Avenue, N.W.
Washington, D.C. 20418

Printed in the United States of America

CONTRIBUTORS

EDGAR BROWN, Department of Mathematics, Brandeis University, Waltham, Massachusetts

GEORGE CARRIER, Deputy Chairman of the Delegation, Department of Mathematics, Harvard University, Cambridge, Massachusetts

WALTER FEIT, Department of Mathematics, Yale University, New Haven, Connecticut

ANNE FITZGERALD, Professional Associate, Committee on Scholarly Communication with the People's Republic of China, 2101 Constitution Avenue, N.W., Washington, D.C. 20418

JOSEPH KELLER, Courant Institute, New York, New York

VICTOR KLEE, Department of Mathematics, University of Washington, Seattle, Washington

JOSEPH KOHN, Department of Mathematics, Princeton University, Princeton, New Jersey

CARL LEBAN, Department of Oriental Languages, University of Kansas, Lawrence, Kansas

SAUNDERS MAC LANE, Chairman of the Delegation, Department of Mathematics, University of Chicago, Chicago, Illinois

HENRY POLLAK, Mathematics and Statistics Research Center, Bell Telephone Laboratories, Murray Hill, New Jersey

H.H. WU, Department of Mathematics, University of California, Berkeley, California

PREFACE

A Pure and Applied Mathematics Delegation consisting of nine mathematicians, an orientalist, and a staff member of the Committee on Scholarly Communication with the People's Republic of China (CSCPRC) visited China from May 3 to May 27, 1976. It was organized under the auspices of the CSCPRC, which is jointly sponsored by the National Academy of Sciences, the Social Science Research Council, and the American Council of Learned Societies. This was the first of four American academic delegations to visit China in 1976 under the CSCPRC's current exchange program with the Scientific and Technical Association of the People's Republic of China (STAPRC). Since 1972, when the exchange program was initiated through the Shanghai Communiqué, the CSCPRC has sent to China 18 delegations in a wide variety of fields and has hosted 25 Chinese delegations. Although this was the first exchange in mathematics sponsored by the CSCPRC, at least 15 American mathematicians and 14 computer scientists have visited China independently since 1972. Through these visits a foundation is being laid for continuing scientific and cultural interaction between Americans and Chinese. The Mathematics Delegation hopes that its visit will be followed by continuing communication with Chinese mathematicians.

In Peking the delegation visited the Institute of Mathematics, the Institute of Meteorology, Peking University, and Tsinghua University and held discussions in the Peking Hotel with scientists from the Institute of Mechanics, the Institute of Oceanography in Tsingtao, the Institute of Biophysics, and the Institute of Mathematics. It also visited a diesel locomotive depot, a printing machinery plant, and a car insulation depot to hear about applications of mathematics. Part of the delegation visited the city of Harbin and the nearby Tach'ing oil field to learn more about applications of mathematics in industry, while the remainder stayed at the Institute of Mathematics in Peking. Then the entire group visited Soochow for a day of sightseeing and proceeded to Shanghai for visits to Futan University and the Shanghai Hua-tung Normal University.

After an opening briefing at each institution, the delegation divided into small groups. On many occasions, subgroups visited separate institutions simultaneously. More than 60 lectures were given to the group during its stay, and its members gave 20 lectures.

The planning by the host organization, the Scientific and Technical Association, was excellent. The staff coordinator for this visit, Su Feng-lin, was extremely helpful in making adjustments in the itinerary to suit individual needs. In addition to him, several members of the Institute of Mathematics accompanied the delegation as interpreters: the official interpreters Chu Shih-hsüeh and Chu P'ei-hua, the operations research specialist Kuei Hsiang-yün, and the differential geometer Tai Hsin-sheng. All of them deserve great credit for their untiring efforts on the group's behalf.

We were received warmly everywhere we went. In addition to our many professional visits, we were taken to see interesting sights and were invited to superb banquets by our hosts in each city. Chou P'ei-yüan, Vice-Chairman of the Scientific and Technical Association, and Liu Hua-ch'ing, Leading Member of the Chinese Academy of Sciences, welcomed us at a banquet in Peking, and Hua Lo-keng, Director of the Institute of Mathematics, invited us to luncheon in the beautiful setting of the Summer Palace near Peking. We were received at the Great Hall of the People by Yao Lien-wei, Vice Chairman of the Standing Committee of the National People's Congress, on our last day in Peking. In every city we were welcomed by many helpful and hospitable people. Our thanks extend to all of our hosts for the full professional, cultural, and social programs they arranged for us.

The report that follows describes some of the delegation's observations and presents its appraisal of the state of pure and applied mathematics and of mathematical education in China.

CONTENTS

1

INTRODUCTION

Mathematics is an international language. Far removed from particular cultural circumstances and political considerations, basic mathematical concepts are universal and most mathematical notations are the same in all languages. Hence, mathematics provides an effective topic for the development of international exchange and for the establishment of common interests between scientists from different countries. It was the mission of the Pure and Applied Mathematics Delegation to establish meaningful contacts with Chinese mathematicians and to learn as much as possible about the present condition of mathematical research, applications of mathematics, and mathematical education in China.

Though the substance of mathematics is international, the state of its development at any time and place depends on local social and economic conditions. In China the indigenous development of mathematics began very early, tapered off noticeably in the fourteenth century, and only gradually revived in the twentieth century. Then the founding of the People's Republic of China in 1949 was followed by a period of increased activity--which was interrupted, however, by the start of the Cultural Revolution in 1966. All universities were closed until 1970, while a new policy was developed for research and training in all fields of science, including mathematics. This policy is guided by the slogans "science must serve society" and "education must combine theory with practice," which mean that research should focus on practical problems and that teaching should be based upon concrete applications.

Accordingly, in mathematics the stress is on applications both in research and in teaching. For example, China has made substantial progress in computer science and has been able to reach the forefront of such areas as queuing theory quite rapidly. In classical applied mathematics, the theory of the finite element method was discovered independently. However, most of the applied research is in the nature of engineering mathematics and its applications.

Despite the current emphasis on applications, the mathematicians who were trained before the Cultural Revolution still manage to devote some of their time to research and study in pure mathematics. Much of the research in pure mathematics is first-rate, and some of the recent results are important contributions. Particular mention should be made of work on the Goldbach conjecture, on the Waring problem, and in Nevanlinna theory. There have also been substantial contributions to modern algebraic topology.

Solid work is being done on a number of other topics, but mathematical work in China is hampered by the isolation of Chinese mathematicians and by the lack of new blood in the profession. The current university course in mathematics usually lasts only 3 years and there is up to now no formal graduate work in mathematics. Therefore, mathematics education does not reach a very advanced level. All these facts are made evident in the detailed reports of lectures, curricula, and discussions contained in this report.

Since we visited only one mathematics research institute, four universities, and about a dozen factories, our observations and conclusions are based upon a relatively small sample. Judging by published work, we probably saw most of the best centers of research in pure mathematics. However, since applied mathematics is much more widespread, what we saw was only a small fraction of it. In universities, we visited only mathematics departments and, as a result, saw very little in related fields such as computer science and information theory. We had only minimal exposure to secondary education. The factories we saw were all operated locally and independently, and we observed many applications of elementary statistics and operations research, and some applications of classical mathematical physics. But we were not shown applications of mathematics on a national scale, as in telecommunications, mathematical economics, nationwide pricing or planning, or military or nuclear energy applications. These limitations should be kept in mind in reading this report.

A Note on the Transcription of Chinese Names

For the sake of consistency, Chinese personal names in this report are uniformly rendered whenever possible into the commonly used Wade-Giles romanization. Place names are rendered into the commonly used "Chinese Postal System" transcriptions.

Appended to this report is a list of the names of our Chinese hosts in Wade-Giles as well as in Chinese characters.

2

MATHEMATICS IN CHINA UP TO 1966

Mathematics developed very early in China. From the third century B.C. to the fourteenth century A.D., the Chinese made many important contributions to mathematics; some of them anticipated the Arabs, the Indians, and the Europeans by several centuries [see (1) at the end of this chapter.] However, around the fourteenth century, mathematics in China became stagnant, until Western science and mathematics were brought to China by the Jesuits in the seventeenth century. A period of activity followed, without contact with the major developments in the West. In terms of the number of competent mathematicians and the availability of advanced mathematical instruction in the modern sense, conditions remained quite primitive as late as 1930. In the thirties, a small number of Chinese mathematicians, fresh from earning their doctorates abroad, began to assume teaching posts. The general level of mathematical education improved. For example, the geometer Su Pu-ch'ing and the analyst Ch'en Chien-kung, both trained in Japan and both teaching at Chekiang University, were forming schools at that time. Although the Sino-Japanese War of 1937-1945 disrupted this development of mathematics, a few Chinese mathematicians continued their work and left their mark on world mathematics--among them S. S. Chern, W. L. Chow, P. L. Hsu, Hua Lo-keng, and C. C. Lin. But all Chinese mathematicians who were trained before 1946 and later gained international reputations spent part of their formative years in the West.

The first serious effort to bring *modern* mathematics to China was made in 1946, the year S. S. Chern was entrusted with the creation of the Mathematics Institute in Nanking. The most talented college graduates in mathematics from all over the country were brought to the Institute and were given intensive training. Due to the disruptions of the civil war, this program lasted only 2 years, from 1946 to 1948, yet it left an indelible impression on Chinese mathematics. Most of the leading Chinese mathematicians who were educated during that period were affiliated with the Institute. They include S. C. Chang, K. T. Ch'en, S. T. Hu, H. C. Lee, S. D. Liao, H. C. Wang, Wu Wen-tsün, and C. T. Yang. (All of them obtained their doctorates in the West.)

The period between Liberation (1949) and the beginning of the Cultural Revolution (1966) was the formative period in China's progress towards mathematical independence and maturity. China was more active mathematically in these 17 years than in any previous such period of its history.

The 10 years from 1949 to 1959 saw the emergence of 342 mathematicians who published 983 papers, as compared with a total of only 74 mathematicians during all the years before Liberation, with only 652 publications, most of them in foreign journals [(2), p. 1219]. Since Liberation, Chinese mathematicians, with few exceptions, have published all their papers in Chinese journals.

It is difficult to evaluate the publications of 1949-1966 because most of the Chinese journals of this period are not available in the United States. Tsao's bibliography of Chinese Mathematics from 1949 to 1960 (5) lists 1,271 titles. These titles give the impression that mathematical activity was centered on technical and sharply defined problems of a more classical nature than those in the West. The mathematical papers in *Scientia Sinica* and in the 15 volumes of *Acta Mathematica Sinica*, the principal mathematics journal of China during the period 1949-1966, essentially substantiate this impression. Moreover, the higher quality work was done by a few people (3),(4). The strongest fields in this period were analytic number theory and algebraic topology. Examples of outstanding achievements in these areas are Wu Wen-tsün's theory of imbedding of polytopes (6) and Chen Ching-jun's best results to date on the Goldbach conjecture (7). In other areas, the research monograph of Hua Lo-keng in several complex variables (8) and his joint monograph with Wan Che-hsien in classical groups (9) are highly regarded. Special mention should be made of the independent discovery of the finite element method by Feng K'ang in 1965 (10); this work is overlooked in the West, probably because it was published in a relatively new journal that has never been translated into English. The differential geometers were also very active (3), although they were heading in the direction of generalized spaces not currently in fashion in the West.

The development of applied mathematics in China was hampered by the absence of a tradition. For instance, while the Chinese realized that a good knowledge of the qualitative theory of ordinary differential equations was important for their technology, they conceded that much more remained to be done to stimulate research in this area (2). In the reports of 1959 [(2),(11)], the authors repeatedly stressed that "mathematical theory should be integrated with practice, and that priority should be given to rendering service to the socialist reconstruction of the new China." It appears that these exhortations succeeded chiefly in inspiring research on a very practical level rather than a more theoretical study of basic phenomena. The paper by Hua on "Applications of mathematical methods to wheat harvesting" (12) may serve to give an idea of this trend. This attitude toward applied mathematics became the dominant theme in all things mathematical after the Cultural Revolution.

To counterbalance the preceding comments, one has to recall the status of mathematics in China in 1949. The road from mathematical awakening to mathematical maturity is always long and difficult; witness the corresponding examples of the United States and Japan early in this century. In the case of the Chinese, there was the additional burden of social reconstruction. Nevertheless, seen from the available evidence, the achievements of the Chinese mathematicians from 1949 to 1966 were altogether creditable. The level of technical competence in the papers of *Acta Mathematica Sinica* at that time reveal a healthy increase in mathematical activity, both in quantity and in quality. The presence of only a few outstanding mathematicians is entirely natural for any development in its early stages. Hua expressed this eloquently in 1959: "Ten years is a short period, and much time has been spent on restoration and rehabilitation, on reformation of our methods of thought, on correction of our teaching methods, and on the training of many staff members, so that our results to date can only be considered as preliminary. However, in comparison with the period before the Liberation, our confidence has been doubled." (11)

REFERENCES

1. Needham, Joseph. 1959. *Science and Civilization in China*, Vol. 3. Cambridge University Press, pp. 1-168.
2. Mathematics Group. Research works in mathematics in China from 1949 to 1959. Sci. Sin. VIII(11):1218-1228. 1959. (English.)
3. Su Pu-ch'ing and Ku Ch'ao-hao. 1960. The development of differential geometry in China for the past ten years. Sci. Sin. VIII(11): 1238-1242. (English.) Republished in Notices Amer. Math. Soc. (7): 163-168.
4. K. H. Look. 1960. A study of the theory of functions of several complex variables in China during the last decade. Sci. Sin. VIII (11):1229-1237. (English.) Republished in Notices Amer. Math. Soc. 7:155-163.
5. C. K. Ts'ao. 1961. *Bibliography of Mathematics Published in Communist China during the Period 1949-1960*. American Mathematical Society.
6. Wu Wen-tsün. 1965. *A Theory of Imbedding, Immersion and Isotopy of Polytopes in a Euclidean Space*. Peking. (English.)
7. Ch'en Ching-jun. Research announcements. Kexue Tungbao 17:385-386. (Chinese.)
8. Hua Lo-keng. 1957. *Harmonic Analysis of Functions of Several Complex Variables in the Classical Domains*. Peking. (Chinese.) English translation by the American Mathematical Society.
9. Hua Lo-keng and Wan Che-hsien. 1963. *Classical Groups*. Shanghai. (Chinese.)
10. Feng K'ang. 1965. Differencing schemes based on the variational principle. Appl. Math. and Computational Math. 2:238-262. (Chinese.)
11. Hua Lo-keng. 1959. Mathematical research in China in the last ten years. Sci. Sin. 18:565-567. (Chinese.) English translation in Notices Amer. Math. Soc. 6:724-730.
12. Hua Lo-keng. 1961. Applications of mathematical methods to wheat harvesting. Acta Math. Sin. 11:77-91. (Chinese.) English translation in "Chinese Mathematics," published by the American Mathematical Society.

3

"POLITICS IN COMMAND": THE POLITICAL SETTING FOR CHINESE MATHEMATICAL ACTIVITY

At each visit to a university or institute, we were given an introductory speech by a vice-chairman or a responsible member of the Revolutionary Committee. Before presenting basic facts about the institution, the speaker would invariably refer to the campaign against Teng Hsiao-p'ing, which was current at that time, and the struggle to combat the "right deviationist wind" or revisionism. The activities of the institution, we were told, are designed to meet the needs of socialist reconstruction, and research is carried on in the "open-door way." Scientists must learn from the workers, peasants, and soldiers.

In short, before opening the mathematical discussions, our Chinese hosts were always careful to present the political and social considerations that affect their mathematical work. (For an example, see Appendix D.1). In this spirit, we will summarize, to the best of our understanding, how such considerations may influence the shape of mathematics in China.

THE GOALS OF SCIENTIFIC RESEARCH

The Marxist theory of the history of human knowledge holds that theoretical study is derived from economic necessity and that there is an intimate interplay between the development of science and its application to production. This theory has been interpreted in China, since the Cultural Revolution, to mean that theoretical work should not be pursued for its intrinsic value but rather in connection with immediate applications. Policies resulting from the Cultural Revolution focus on two points: science must serve society and education must combine theory with practice. In other words, research should be directed toward practical problems, and teaching should be based upon concrete applications.

One of China's major goals is to become a self-sufficient and independent industrialized nation. Increasing production is an important aspect of China's scientific policies. To a Westerner, this would seem to require a vigorous program of education and research in theoretical and applied science and in mathematics. However, a second and equally important aim in the minds of the Chinese leaders is to shape the political consciousness of the participating scientists. While stimulating the economic and technological development of the country, they are trying to avoid the rise of an elitist group of technocrats. The problems described below have been the source of much political struggle in recent years.

> Prior to the Great Proletarian Cultural Revolution. . . . because of the interference and sabotage by Liu Shao-chi'i's revisionist line, some areas of research were dominated by a small number of bourgeois intellectuals and scientific research was for a long period done behind closed doors completely cut off from the workers and peasants. Some of its personnel did not take into account the country's interests, one-sidedly stressed personal preference and the "free choice of subjects" and went about setting up self-centred "bases." Some who were bent on establishing their reputation engaged in working on "theoretical systems" behind closed doors. Some even corrupted young scientific workers with bourgeois ideas of seeking personal fame and gain. This led a number of people to devote themselves entirely to acquiring knowledge as personal capital and neglect the tasks assigned to them, thus gradually embarking on the revisionist road. All this seriously affected the socialist orientation of scientific research.(1)

Ever since the founding of the People's Republic in 1949, there has been a continuing philosophical debate over priorities: whether ideology or economic and technological development should be paramount. Characteristic of post-Cultural Revolution China is a renewed emphasis on ideology. Scientific institutes and universities can no longer concern themselves exclusively with scientific research; they must engage in a continuing political struggle to prevent "revisionist" tendencies. Chinese mathematicians spend a substantial amount of time and energy on activities whose principal aim is political. At the Institute of Mathematics, for instance, at least two afternoons a week or all day Saturday are spent in political study and discussion.

To reinforce the "revolutionary line" on scientific development, scientific institutes and universities have engaged since the Cultural Revolution in two practices that are constantly mentioned: "open-door" research and the "3 in 1" method.

"OPEN-DOOR" RESEARCH AND THE "3 IN 1" METHOD

According to common description, "open-door" scientific research is based on the philosophy that "scientific research should serve proletarian politics, serve the workers, peasants, and soldiers, and be integrated with productive labor." This approach means that the doors of intellectual institutions should be open to the problems of production. Professors, research workers, and students of the universities and the staff of the institutes are regularly assigned tasks in (sometimes distant) factories or communes; peasants and workers are invited to institutes and universities to take part in research work; and research institutions cooperate with other organizations in carrying on research projects. This system is intended to establish a controlled interchange of information and personnel among universities, research institutes, factories, and communes, thereby helping to establish and consolidate the dictatorship of the proletariat.

The impact of the open-door method on mathematics is manifest in the overriding concern of Chinese mathematicians with problems immediately applicable to production. The obligation of all scientists to participate in the open-door program seems to have reduced greatly the possibilities for pure research. For instance, all mathematics students must spend part of their academic year in production, such as manufacturing computing machines. Scientists may spend over a year in the field, working out solutions to local problems. As one mathematician at the Institute explained, "Finding topics for research from the literature has been completely changed since the Cultural Revolution. . . . As a mathematician, you must learn from and integrate yourself with the workers, peasants, and soldiers. You must use their language, simplify, select the essentials, make it easy for the workers, peasants, and soldiers to understand." What is required is a change in attitude and a change in work style.

> In China, a socialist state under the dictatorship of the proletariat, the workers and peasants are the main force in the three great revolutionary movements of class struggle, the struggle for production and scientific experiment. Science and technology cannot develop if they are divorced from the workers and peasants and deprived of their rich practical experience. By carrying out open-door scientific research and taking the path of integrating with the workers and peasants, scientific and technical personnel have the opportunity of being politically re-educated by the workers and peasants, and this helps them remould their world outlook. At the same time they can restudy what they have learnt and integrate their book knowledge with the practice of production. This accelerates the development of science and technology. (2)

Side by side with the open-door policy is the "3 in 1" principle, which holds that the proper way to do research is by combining the efforts of cadres (administrators), technicians (mathematicians, in our case), and workers. The activities of the technicians and workers can be imagined easily, but the role of the cadre may require additional explanation for a foreigner. Aside from "assuring that the work will have a proper political orientation," the cadre is useful in making the physical arrangements for the visits of scientists to factories, in helping scientists and workers to communicate, in encouraging their joint efforts, and in convincing the managers of the factory to permit the experiments necessary for improving production. (This is not trivial, for it often involves interrupting production.)

"3 in 1" also refers to teamwork among the old, the middle-aged, and the young. In effect, the open-door and "3 in 1" methods are ways to accelerate the popularization of science and to involve people of all ages and backgrounds in China's drive toward technological and economic development.

GENERAL CHARACTERISTICS OF CHINESE SOCIETY THAT BEAR ON SCIENTIFIC WORK

A number of themes presented themselves so often in China that we became interested in their impact on mathematical activities. To begin with, the constant emphasis on unselfishness and "serving the people," which is reinforced by the open-door method, has served to re-orient scientific activities. An effort is made to instill in small schoolchildren a pride in helping others, and this effort continues when middle-school graduates are sent for at least 2 years to factories and communes to be of service to and to learn from the workers, peasants, and soldiers. The context in which the system operates is that people are not supposed to express personal preferences, but rather to live by the notion of contributing their efforts where they are needed. With respect to mathematicians, it is hoped that higher esteem will be accorded to those who help factory workers solve problems rather than to those who prove theorems in a research institute. Mathematicians are supposed to think first of benefiting their country and the people's livelihood rather than of pursuing problems that are intellectually stimulating to themselves.

The motivation to "serve the people" is backed by the intense patriotism of the Chinese people. The feeling is widespread that their country was only recently liberated from foreign domination and exploitation, and one frequently hears that their present situation is substantially improved over the recent past. The experiences associated with previous eras of utilizing foreign expertise to help build up their country play a major role in their determination now to develop technology, including mathematical applications, as much as possible by their own efforts. For example, Chinese mathematicians are proud of the fact that they have developed by their own efforts substantial parts of the finite element method for the stress analysis of composite structures. However, this emphasis on self-sufficiency is accompanied by real care in noting all references in the foreign scientific literature that may be relevant to the scientific problem in question.

Another value that no doubt has an effect on the sciences and mathematics is equality, or egalitarianism, which has been given a renewed thrust since the Cultural Revolution. Contrary to the Confucian tradition, scholars are not to be considered superior to manual laborers. To attend a university, a student must be a worker, peasant or soldier, although it seems sufficient to have served for 2 or more years following middle school to qualify as a member of one of these classes. Manual labor is stressed as an admirable occupation. It is also held that there is no such thing as special talent. Our own experience in mathematics leads us to believe that talent is not equally distributed and that unusual talent may need particular encouragement. The Chinese, however, in the interest of egalitarianism, are seeking to avoid such individual recognition. One frequently hears it stated that "everyone can do mathematics."

While the Chinese stress equality in all endeavors, they seem at the same time to be much more subtly conscious of hierarchy than are we from the West. Order of precedence at banquets and receptions is of great importance. In greeting a group, it is important to shake hands first with the correct person. People in responsible positions are treated with great deference.

The pressure on the Chinese people to conform to general expectations leads a foreigner to wonder about the implications for scientific inquiry. Due to the highly organized nature of Chinese society all the way down to the neighborhood level, peer observation and discussion are important components in daily life. Political issues are described in terms of the "conflict between the two lines," currently meaning the conflict between Chairman Mao's revolutionary line and the revisionist line of his opponents. Everyone is expected to know the correct line and to criticize those who deviate from it. The practice of criticizing oneself and one's fellows is central to defining orthodoxy, and mass discussion is the primary mode for transmitting and clarifying the official line.

Such emphasis on consensus is hard for a contemporary Western mind to fathom. We do not know what effect it may have on scientific creativity. Conventional Western thought holds that some vital advances in science depend on "breakthroughs" that often are not accepted at the moment. We wonder how the pressure to conform affects the development of unconventional scientific ideas. In our experience, the progress of mathematics, as of other sciences, depends vitally on people with iconoclastic new ideas. How are such people integrated into the Chinese system? These considerations are subtle and we do not have enough perspective to know how they now apply in China.

REFERENCES

1. Scientia Sinica. March-April 1976. XIX(2):175.
2. Scientia Sinica. March-April 1976. XIX(2):176.

4

RESEARCH INSTITUTES AND UNIVERSITIES

We shall now sketch the structure and mathematical work of some of the institutions we visited or had contacts with: the Institute of Mathematics, Peking University, Tsinghua University, Futan University, Shanghai Hua-tung Normal University, Heilungkiang University, and Harbin Industrial College. This information is based on what we were told in introductory statements by our hosts, who, in each case, went to a great deal of trouble to present us with the background and the basic facts.

THE INSTITUTE OF MATHEMATICS (ACADEMIA SINICA, PEKING)

The Institute of Mathematics, which played a central role in our delegation's visit to Peking, appears to incorporate the major mathematical activities of the Academia Sinica. Planned in 1950, the Institute was founded in 1952 with a staff of about 20. By 1958 it had over 100 research workers and now it has over 200. Until 1966 its research was primarily in pure mathematics and was carried out by seven research groups in number theory, algebra, topology, differential equations, probability and statistics, computer science, and numerical mathematics. During the Cultural Revolution, which began in 1966, its work was reoriented toward applied mathematics.

Hua Lo-keng, the Director of the Mathematical Institute, is not often present since, among many other responsibilities, he works with operations research groups in about 20 different provinces. Chao Wei-shan, not a mathematician but an administrator with military background, is the Chairman of the Revolutionary Committee of the Institute and T'ien Fang-tseng is the Deputy Director of the Institute.

During the last year, 8 or 10 new members (none in pure mathematics) were added, although the Institute had hoped for 20. There is a "planning group" of 12 members, of whom about six are mathematicians. The research workers are organized into seven divisions with 20 to 40 people in each: partial differential equations, probability and statistics, operations research, control theory, five subjects (number theory, algebra, topology, functional analysis, and function theory), computer studies, and a laboratory for calculation. There is a separate Institute for Computing Technology that was once part of the Mathematical Institute.

Members of the Institute occupy the top 1½ floors of a large building in a northwest suburb of Peking, near other institutes of the Academy. The lower floors of the building are used by the Institute of Computing Technology, which also occupies an entire building next door. There is a crowded mathematics library and a reading room for current journals, including about 250 current journals from the West and 20 from the USSR. The research workers at the Institute share their offices, often with four or six in a relatively small room. There are several small lecture rooms and at least one large lecture room that holds about 200 people.

Mathematical Work of the Institute

In pure mathematics, there is considerable emphasis on *number theory*, with activity by several former students of Hua Lo-keng--especially in analytic number theory (the Goldbach conjecture), diophantine approximation, and applications to numerical integration. Research in *algebra* includes the geometry of matrices, classical groups (as represented by the book on this subject by Hua and Wan Che-hsien), finite geometries, and incomplete block designs and other applications to combinatorics. *Geometry* and *topology* include work on new topological invariants (the I^*-functor and other developments by Wu Wen-tsün of Dennis Sullivan's ideas), studies of cohomotopy groups, dynamical systems, and catastrophe theory. In *analysis*, there is special emphasis on value distribution theory for meromorphic functions, especially on Borel directions, in the work of Chang Kuang-hou and Yang Lo. In *functional analysis*, spectral analysis and application to mathematical physics are stressed. Extensive work on *partial differential equations* includes second-order equations of mixed type, nonlinear hyperbolic equations, and other topics.

Mathematicians at the Institute work on applications in agriculture (breeding of seeds, fertilizers, comparisons between species, distribution of labor) and in industry (chemical industry, light industry, metallurgy, electronics, medicine, railroads, construction, etc.)

Much attention is paid to optimization procedures, especially optimal search, PERT, and orthogonal experimental design; every province and city is said to have a special organization to foster this type of work. The *operations research* group now includes about 30 people, many of whom spend time visiting factories. Graphical methods, reliability, quality control, and network optimization problems are emphasized. For example, linear programming is used to minimize the cost of the transportation of grain. Members of the section on *industrial control* study such problems as stochastic observability, weather forecasting, the stability of elastic vibrations, and the use of mathematical models in the control of industrial processes. In the section on *differential equations*, studies of the Laplace equation are related to problems of prospecting. The *computer science* section does the calculations necessary for the work of the Institute. However, the main research in computer science is done at the Institute for Computing Technology. Work on *probability and statistics* includes time-series, experimental design, and sampling questions.

In 1974, the Institute began publishing a new journal, *Mathematics: Its Practice and Cognition*, which has a circulation of 40,000-50,000. This journal is designed to make mathematics comprehensible to prospective users. At the present time, most of the articles are of an applied nature.*

PEKING UNIVERSITY

Peking University, founded in 1898, has 20 departments (seven in the arts, three in the languages, and ten in the sciences) with 75 specialties. Mathematics is one of the 10 science departments. There is a teaching staff of 2,700, and the university operates seven factories and three farms. The impressive new library has 3 million books, of which 800,000 are in foreign languages, and has more than 1,400 foreign journals. The university is located on a beautiful campus (with a small lake, bordered by weeping willows and pavilions) in a suburb of Peking not far from the Mathematical Institute.

In most departments of the university (including mathematics), the students follow a 3-year course. In a few departments (theoretical physics, Arabic languages, and art history) it is a 4-year course, and in library science, 2 years. There is some discussion about starting graduate work, which now exists in only a few departments. The university has about 6,000 regular students and hopes to expand to 10,000 within the next 5 years. About 50,000 students are enrolled in short-term, refresher, and correspondence courses. Short-term courses last from 2 to 5 months, with 40 to 100 students per class. There are also about 150 foreign students from 31 countries, most of whom study Chinese or social science.

After the Cultural Revolution, the university reopened to students in 1970 and 1971. A small group of students was admitted to the computer science department at that time. The first students in mathematics were admitted only in 1973. They will graduate in 1976--the first graduates in mathematics since the Cultural Revolution. Since 1973 about 140 students have been admitted each year in mathematics.

The mathematics department consists of three divisions:

- The mathematics division (pure and applied)
- The computational mathematics division
- The information theory division

*See Appendix E.3.

In the mathematics division, the first-year course, taught by a faculty of six mathematicians (plus teachers in other fields), starts with a review of middle-school mathematics. This is necessary because the students have been working on farms or in factories for 2 or more years since leaving middle school. Then there are some courses in calculus, analytic geometry, physics, English, and philosophy. In the second year, again with a faculty of six mathematicians, the 50 students in this division study advanced calculus, ordinary differential equations, linear algebra, and English. At some point, they also study mechanics and computer programming. In the third year the 50 students are split into three groups. About 20 students study complex variables, calculus of variations, and electricity and magnetism. A second group of 10 students studies Fourier analysis, measure theory, and eigenfunction expansions. The third group of 20 students studies partial-differential equations and the finite element method.

The course on complex variables provides an example of the teaching methods. As noted in the previous paragraph, the students in the course also study electricity and magnetism. The teachers felt that the usual definition of an analytic function $f(z)$ as one having everywhere a first derivative was too abstract; starting with the Cauchy-Riemann equations was considered ideologically incorrect. Instead, analytic functions were introduced in terms of the corresponding irrotation source--free fields in the plane. One of the students in the course gave us the essential points of the lecture introducing this notion. This new approach was greeted with enthusiasm.

TSINGHUA UNIVERSITY

Tsinghua University, founded in 1911, is devoted exclusively to education in engineering and technology. There are 11 departments: electrical engineering, mechanical engineering, automation engineering, radio engineering, precision instruments, electrical power, engineering physics, engineering mechanics, chemical engineering, architecture and civil engineering, and hydraulic engineering. The student body numbers 10,000 and the staff 9,000 (of which 3,500 are faculty). There is a secondary school attached to the university, plus 20,000 students in short courses (evening courses and correspondence courses). The university runs 25 factories and workshops.

The changes in teaching methods since the Cultural Revolution have been enormous. We were told that before the Cultural Revolution, the students crammed for examinations, then threw away their notebooks and textbooks after the examination, and forgot all! Not only are there no examinations now, but there have been almost no formal classes since November 1975. Students are taught largely through working in the university's factories and workshops under the supervision of their professors. At any given time, 50 percent of the students are away from the campus, working in factories, communes, or the army, accompanied by their teachers. The students are faced with new situations in their practical work and are then taught some science in order to raise their ability to deal with these new problems. Last year, the third-year students carried out 560 projects. Some of these projects involved research and others involved practical problem-solving or designing work.

First-year students at the university do not all take the same courses, for each department has its own courses. There is no mathematics department, but each department teaches its own calculus and differential equations, covering the aspects specifically needed in that department. We asked whether this system might not sacrifice a cross-fertilization of ideas between departments. In response, we were told that courses constructed in this way were much more efficient and that each student can study to the extent of his interest and his available time.

Power engineers study complex variables and transforms, in addition to elementary analysis. Some students also study linear algebra and a few study probability. We asked about the theoretical background of the students we observed working on laser assembly. We were told that most of them are engaged only in assembly work but that 10 to 20 postgraduate students work on the theory of lasers. For this purpose, they learn quantum mechanics and differential equations.

Since the Cultural Revolution, the method of education has shifted from an encyclopedic "cramming method" toward methods intended to raise the ability of students to solve new problems after graduation. It is said that the best way to learn to confront new problems is to confront them in school itself. We were told about a film called "Break Away from the Old Ideas" based on an incident that illustrates previous shortcomings in teaching. A teacher of veterinary medicine used to teach his students just about "the function of the horse's tail"--knowledge that is useless for the treatment of a sick cow. This teacher changed his mind after the Cultural Revolution and now teaches about cows, too, as he should. The course is now geared toward practical problems that the students will confront in their lives. We were told that the peasants very much welcomed this change.

SHANGHAI HUA-TUNG NORMAL UNIVERSITY

Founded in 1951, the Shanghai Normal University trains 3,000 to 5,000 middle-school teachers every year and provides in-service training for middle- and primary-school teachers. The teaching staff numbers 2,000 and there are currently 7,000 students living on campus. Since the Cultural Revolution, the Normal University has enrolled other students who are not living on campus--for example, 15,000 correspondence-course students, students from various Shanghai factories who enroll in technical-training courses, and teachers from state farms and communes who enroll in the teacher-training courses.

There are 12 departments: education, political education, Chinese, history, foreign languages (English, Russian, Japanese, German, and French), literature and art, physical culture, mathematics, geography, chemistry, physics, and biology. Attached to the Normal College are three research institutes focusing on education in foreign countries, foreign geography, and estuaries and coasts. There is also a research group for middle-school teaching.

The four factories located on campus are concerned with metal-cutting and electronic equipment, semiconductor elements, microbiology, and chemical works.

In the mathematics department there are seven teaching groups: computational mathematics, applied mathematics, application of computers, training of middle-school teachers, physics teaching, research on mathematics in middle schools, and correspondence courses. There are 185 teachers in the department, one-fourth of whom are women. Seven hundred and fifty-five students are enrolled in the mathematics department: 575 students take a 3-year course and 180 take a 1-year course. Starting in 1974, the 3-year students were divided into three units: computational mathematics, applied mathematics (study and research), and application of computers. Postgraduate students were enrolled last year and this year. Two are studying statistics.

All students at the Normal University are required to take the following courses: political theory (especially Marxism and Mao Tse-tung's teaching on education), a foreign language, Mao Tse-tung Thought, and physical education. Ten additional courses are offered to the mathematics students (mainly the 3-year students): algebra and geometry, calculus, differential equations, algorithms, linear algebra, probability and statistics, finite element method, principles of computers, mechanical drawing, and physics. The amount of time spent on these subjects varies according to the student's specialty.

FUTAN UNIVERSITY

Futan University, founded in 1905, is the leading university in Shanghai. It has seven departments in the liberal arts: Chinese, history, journalism, political science, foreign languages, philosophy, and international politics and affairs. Its seven departments in the natural sciences are mathematics, physics, chemistry, biology, optics, computer science, and atomic energy. Of these, computer science and optics have been added very recently. These departments are subdivided into 41 sections or specialties, 17 of which are new since the Cultural Revolution.

The usual course of study runs for 3 years, plus 2 years of postgraduate work in a few cases. There are also short courses, correspondence courses, and radio and television classes. More than 3,000 regular students are currently enrolled (including 113 postgraduates and 19 foreign students). The number of students taking short-term courses has increased considerably: 6,000 in 1974, 8,000 in 1975, and 10,000 in 1976. Correspondence courses are given for middle-school graduates who have gone to work in rural areas in four specified provinces. The number of these correspondence students was 6,000 in 1974, 16,000 in 1975, and 18,000 in 1976. All told, the university now has more than 30,000 students.

Futan University runs five factories on campus and has permanent links with at least 120 other factories and communes. Three of the factories on campus work on electrical instruments, chemical products from oil, and optical instruments, respectively. There are also 10 workshops on campus. Five institutes are attached to the university: mathematics, genetics, economics, Chinese language, and historical geography. The mathematics institute, which is affiliated with the mathematics department, consists of six faculty members of that department, including Gu Ch'ao-hao and Hsia Tao-hsing. Members of the Institute are given a reduced teaching load.

The staff of the university totals 4,000, of whom about 2,000 are teachers. This number has increased by 50 percent since the days before the Cultural Revolution. At the time of the Cultural Revolution, workers and peasants were invited to the university. "After this revolution, the younger and middle-aged teachers made much progress and some of the older professors also made progress. The university persists in linking education with productive labor and in carrying on open-door schooling."

Our hosts went on to say, "The students of natural science carry out their studies so as to maintain a link with typical practical scientific problems. In the last 2 years, the university has carried out over 200 scientific research projects; of these, 95 percent have been done with student participation."

"In academic research," they said, "the university follows Chairman Mao's policy: 'Let One Hundred Flowers Bloom.' In modern physics, for example, there were over 200 discussion sessions in all departments so as to help the students to study dialectical materialism and to realize the leadership of the working class."

After this general presentation, our discussions centered on more specific questions. We learned, for instance, that the postgraduate program is still very experimental. It is expected to develop gradually. The division of students by department is roughly as follows:

Physics	500
Chemistry	300-400
Mathematics	over 200
Atomic Energy	over 200
Biology	over 200
Optics	over 300
Computer Science	over 500

Next, a spokesman for the mathematics department described its activities. He started by observing that before the Cultural Revolution, the mathematics department was revisionist, and its leadership was not closely associated with the proletariat. In 1968, during the Cultural Revolution, a Mao Tse-tung propaganda team of workers marched into the mathematics department. Under party leadership, the students and teachers criticized revisionist policies such as those implied by "knowledge is my own" or "I am learning in order to become an official."

In 1970, new students were enrolled in the university according to the needs of the country. The students came from the workers, peasants, and soldiers (meaning that they had worked in factories, communes, or the army before coming to the university).

Now in the various sections of the mathematics department there is a basis for an open-door policy. The students have solved more than

200 practical problems, more than 100 of them linked with teaching. Most have been put into use, for example, at a Shanghai instrument plant and in Shanghai automobile factories. The mathematics department has managed more than 35 short-term classes with more than 2,000 students. Some of these classes run for several months and some for a year. For example, there have been training classes in the finite element method and in ALGOL. In addition, there are correspondence courses in mathematics for over 600 students. Several of the university mathematics teachers have gone for 1 or 2 years to teach in Tibet.

An area that has developed rapidly is computer science, which was established in August 1975 as a separate department with three sections (previously part of the mathematics department): computational mathematics, software, and information theory. The mathematics department now has four sections: mathematics, mechanics, a fundamental-course research group, and a mechanics laboratory. The staff of the mathematics department numbers 144, of whom 111 are teachers.

The first section, mathematics, was started in 1973. The teaching is organized around three topics: curves and surfaces in industry, the finite element method, and mathematical methods in industrial process control. The chief orientation is toward applying mathematics and using new teaching materials. For example, last year some second-year students went to the oil fields for open-door schooling. They studied electrical phenomena associated with exploration for oil. The teaching for these students was organized around this project for 8 months and included topics such as linear algebra, the finite element method, and the equations of mathematical physics. This kind of teaching involves four steps: preparatory study of electromagnetism and linear algebra, a month spent in the oil fields, further study of mathematical methods required for the problem, and a complete solution of the problem. Another part of the teaching was devoted to advanced calculus and electromagnetic field theory.

DISCUSSION OF EDUCATION AT HEILUNGKIANG UNIVERSITY AND HARBIN INDUSTRIAL COLLEGE

Heilungkiang University, founded in 1958, has 2,000 students following a 3½-year course. The university staff numbers 1,000, including cadres, political leaders, and administrators. Mathematics is one of nine departments and has a faculty of 90. Eighty to 90 students enroll in this department each year.

There are three different programs in the mathematics department: mathematics, numerical methods, and computer science. The 20-30 students in the mathematics subgroup take the following courses in their specialty:

1st Year	Higher algebra, calculus, analytic geometry

2nd Year	Ordinary and partial differential equations, theory of probability, complex variables*
3rd Year	Graduation project connected with practical work

Usually after 1½ years at the university, four to five students begin a cooperative project. A project may be selected by teachers, cadres, and/or students, who go to a factory with a previous idea of what they would like to do, or else a factory may mention a problem on which it would like help. Some projects are chosen by the state.†

There is a very high faculty-student ratio, but in addition to teaching, the faculty engages in research (always connected with practical problems) and also in the time-consuming program of taking students to the countryside or factories for their practical work. Teachers must often go first, to explore the possibilities for projects and to prepare for a specific project. Sometimes teachers are enlisted by the city of Harbin to accompany middle-school students to the countryside--often for a year. In another case, a mathematics teacher may be asked to teach a physics course one year. (The number of teachers in the mathematics department--90--includes those teaching physics and those teaching English to the mathematics students.) The university is reluctant to take on more students until it has codified its educational program.

Like Tsinghua University in Peking, Harbin Industrial College provides training in engineering. There is no mathematics department, but there are mathematics teachers in other departments, whose jobs are to teach only the basic theories of mathematics that are connected with those specialties. All students learn calculus, but the amount of time allocated depends on the specialty of the student. The same is true for ordinary differential equations. Some students may take 150 hours of calculus, and halfway through, begin an 80-90-hour program in linear algebra. Probability is taught to students in the computer science division. The radio and electronics students study the Laplace equation, but without the theoretical background.

There are 11 departments at the Industrial College, but they are in the process of reorganization. Students have no textbooks but use duplicated materials. We were told that the program of the college is still unsettled.

During their 3 years at the Industrial College, students spend 3 months working in a factory, 1 month working on a commune, 1 month in the army, and 3 months in society studying the political struggle. For example, this year, at the invitation of the government of Heilungkiang Province, students from both the Industrial College and Heilungkiang University spent 3 months in the financial department for political study.

*Some of these courses may continue into the third year.

†One project currently under way is in a state-run paper mill, where a computer is used to control the heating of pulp in a vat. The control temperature, the pressure, and the amount of alkali are varied to optimize the production rate of paper.

We were interested to know how educational institutions evaluate the proportion of time spent studying and time spent in production. "We do it by 'summing up experiences,'" we were told. "We cannot use mathematical methods to analyze this!"

We wondered how they evaluate their whole experiment in education. "The policy and the line are not experimental anymore. The general principles have been set. The problem is how to improve teaching materials," we were told. The principles that have been established are the enrollment of students from the workers, peasants, and soldiers and the use of the open-door method, which helps students and teachers to remold their thinking and to apply their knowledge. The assignment of teachers and students to work in factories is considered a fixed principle, but the details are flexible. The optimum length of stay, the timing, and the kinds of projects are still matters under discussion.

"If a university decides that it needs a substantial change, can it go ahead and make the change?" we asked. In general, we were told all universities must follow the same guidelines, as put forth by the Peking Bureau of Education, but the details are different at the various universities. "If one school decides that studying complex variables is not as valuable as doing more production work," we asked, "can it drop this course? And how?" Our Chinese colleagues in Harbin had not experienced such a case. But if a university should want to cancel a course or to initiate something new, these ideas would be discussed by the students, faculty, and cadres in the department. The Party decides at the department level and the plan would then go to the university authorities for approval.

We asked if any student ever desired to transfer from one curriculum to another. We were told of the case of one girl in mathematics who asked for such a transfer. The change of departments was refused. When the other students found out about the request, they all pitched in and helped her, and she did all right.

"Have there been any suggestions to increase the number of courses and decrease the amount of time spent in production outside the university?" we asked. There are such cases. For example, in working on a practical project, the teachers may discover that more theory needs to be learned; they can then make ad hoc additions to the course/project content. (The courses are thoroughly related to the projects, so they cannot be separated conceptually.) So far, there has been no case of adding major new courses to the curriculum at Heilungkiang University or the Harbin Industrial College. But within each course, it is natural to add or subtract topics of study.

Of course, lots of disagreement occurs in the faculty-student-cadre meetings. Some people feel that they ought to teach one subject and others feel they should teach another. But once it has been decided, everyone must adhere to the decision.

5

PRESENT STATUS OF MATHEMATICAL RESEARCH

5.1. APPLIED MATHEMATICS

Classical Applied Mathematics

Since the Cultural Revolution, mathematics, like most science in China, has been re-oriented to "serve production" and to "serve society." Accordingly, mathematics is now taught from an applied point of view and research is now directed toward practical problems.

This policy constitutes a nearly complete reversal of the earlier Chinese attitude, in which attention had been directed almost exclusively to pure mathematics. There was no tradition of research in applied mathematics on which to build, and there were practically no experts to lead in this new direction. The consequences are evident in the present state of applied mathematics.

First, there appears to be little research on the invention and development of new techniques for solving differential and integral equations, on novel variations of existing techniques, or on the fundamental problems associated with such techniques. These activities constitute an essential part of classical applied mathematics in the West and are of great importance for science and technology. *Second,* there seems to be no general awareness of the great variety of methods of classical applied mathematics as developed in other parts of the world during the last 30 years. These include such methods as singular perturbation theory, boundary layer analysis, ray methods, the method of matched asymptotic expansions, the two-time and multi-scale techniques, turning point theory, and the method of uniform asymptotic expansions. Although these methods dominate current Western work in classical applied mathematics, they seem to be completely absent from the Chinese work we saw.* *Third,* applied mathematics in China is

*We remind the reader that what we saw may represent only a small fraction of the Chinese effort. Indeed, from scientific publications (e.g., current issues of *Scientia Sinica*) and other sources of information, it appears that among the scientists and engineers there may be more competence in classical applied mathematics than we saw. Perhaps we failed to meet these people because they are not members of institutions of instruction and research in mathematics.

generally directed at very specific technological problems and is not addressed to the general properties of the solutions of large classes of problems. Instead it seems to aim for the particular solutions of particular problems, often found by numerical methods using electronic computers. This gives much of the work the character of engineering analysis rather than that of applied mathematics. *Fourth,* many of the older standard techniques of applied mathematics are employed. These include such techniques as separation of variables, Fourier and Laplace transforms, and eigenfunction expansions. When these methods are inapplicable, there is no resort to modern analytic methods. Instead numerical methods are employed on electronic computers. Thus, either exact, explicit solutions are obtained for simple problems or numerical solutions are obtained for particular complex problems.

The foregoing account makes it clear that electronic computers play a large role in Chinese applied mathematics. One might expect a good deal of activity in numerical analysis, but such is not the case, with the important exception of the finite element method. This method is widely used and studied. In fact, it was developed independently in the early 1960's by Feng K'ang. (He has recently devised a theory of elliptic partial differential equations on domains consisting of manifolds of different dimensions joined together, in order to analyze complex elastic structures, and he has extended the finite element method to such equations.) To expedite the use of the finite element method, a machine procedure for triangulating a plane domain has been devised. To treat singularities, a modification of the finite element method using an infinite number of similar elements has been developed.

More disturbing than the absence of an influence of modern applied mathematics is the absence of any educational effort that might rectify this discrepancy. There seems to be no systematic attempt to educate young people along lines that would provide a reservoir of skill and understanding in those modern techniques and in the natural phenomena involved, such as boundary layer flow and wave diffraction. Such a reservoir of skill would inevitably become an enormously important resource to China as its industrial stance becomes more and more sophisticated. The gap between activities in mathematics per se and routine applications of mathematics to real-world problems, however cleverly carried out, should not be allowed to remain.

Operations Research

Training in elementary methods of operations research (OR) occurs in some middle schools and in short courses given at factories. University work in OR is carried on in mathematics or specific OR departments. However, the Division of Operations Research of the Institute of Mathematics of the Chinese Academy of Sciences seems to have prime responsibility for developing new methods of OR, recognizing the applicability of known methods to problems in industry and agriculture, and disseminating information. Our report is based mainly on contacts with the Division of Operations Research and with industrial and agricultural users of OR.

The Division currently has about 30 members, of whom usually seven to 10 are away on consultation. The Division's work is strongly influenced by Chairman Mao's advice to "combine theory with practice" and to "avoid the three divorces" (from politics, from the workers and peasants, and from practice). Members of the Division travel extensively throughout the country, lecturing and giving advice on problems. Especially active in this work has been Hua Lo-keng, Director of the Institute of Mathematics, who is internationally known for his earlier work in analytic number theory. He has visited all sections of the country, and once lectured (by telephone network) on optimization to an audience of 100,000. He has been a direct consultant to a number of large projects (for example, the oil and petrochemical industry in Tach'ing, and the forestry operation in Heilungkiang Province) and an indirect consultant, through his protégés, to many smaller ones. In one province alone, more than 5,000 people currently serve on teams devoted to the spread of OR methods.

New members come to the Division with modest university training in mathematics, engineering, or operations research (see Chapter 7.1). Further training is obtained at the Institute, through individual study, seminars, and personal guidance from older members of the Division; however, with the recent hiatus in education, the Division has received almost no new members since the Cultural Revolution. Furthermore, the graduates now becoming available will probably be more immediately useful in the Division's consulting work than in developing new OR methods.

In recent years (1971-1974) the Division's efforts have been concerned mainly with popularizing three elementary but important methods of operations research, which are referred to as OSM (optimal seeking method), CPM (critical path method), and the method of orthogonal experiments. All of these have been applied to many practical problems--for example, Hua Lo-keng estimated that OSM had been applied to more than 100,000 production problems (not all by the Division) since the Cultural Revolution. OSM includes Kiefer's "Fibonacci search" for optimization of a unimodal function of a single variable, and the closely related "golden section search." (Interesting mathematical problems arise when several experiments can be carried out simultaneously, and these have been definitively treated in a paper by Hung Chia-wei, *Scientia Sinica* 17 [1974], pp. 160-180.) An advantage of OSM is that it can be explained to workers unfamiliar with algebra. OSM thus fits well with the Chinese doctrine that workers are capable of almost any achievement. In some of the uses of OSM that we encountered, polynomial interpolation would almost certainly have been more efficient, though mathematically somewhat more demanding. However, great progress has undoubtedly been made through the use of OSM, and perhaps promulgation of the idea of systematic optimization has been even more important than OSM itself.

The "method of orthogonal experiments" refers to the use of block designs in a standard way. The CPM approach to production problems involves formation of a directed graph showing the duration of each task in a project and the precedence of the tasks. Analysis of this graph, sometimes called a PERT chart (for "project evaluation review technique"), enables one to determine such items as the time required to complete the entire project and the latest time at which a given task may be started without delaying the completion time. Particular sequences of tasks ("critical paths"), whose timely completion is essential, invite the assignment of more workers with the aim of reducing the overall completion time. We saw effective applications of this technique in a diesel locomotive repair plant, and we were told in detail about applications to a large logging operation and in an agricultural commune. (Some of the applications involved mutual help by workers in ways that might not be possible in other countries.)

The Division's work is currently distributed as follows: 10-20 percent explanation and popularization of elementary methods--smaller than in recent years; 25-30 percent study of known methods, for better understanding and possible applicability to problems in China; the remainder, improvement of existing methods and developments of new methods.

Though we heard in detail only about application of the three elementary methods mentioned above, more sophisticated methods have been applied in some instances, and were well understood by members of the Division. Mention was made of the simplex algorithm (using the Bartels-Golub approach to reduce round-off errors), the Dantzig-Wolfe decomposition procedure, the Gilmore-Gomory approach to the stock cutting problem, branch-and-bound methods applied to integer programming problems, steepest-descent techniques of optimization, various direct (gradient-free) methods of unconstrained optimization, the Eaves-Kuhn-Scarf method of approximating fixed points, etc. The following were described as the main areas in which members of the Division are currently working: OSM; mathematical programming (including critical path method, simplex algorithm, network programming); stochastic optimization (including queuing problems, Markov decision problems, and simulation); operations research and quality control; and mathematical economics.

In summarizing our impressions of the state of operations research in China, we focus on two opposite aspects:

1. Popularization of known elementary methods, and consultation on specific practical problems
2. Development of new methods

We believe that (1) has been done very well, and that much credit for this should go to Hua and his colleagues for their inspirational efforts. Concerning (2), we are uncertain. Recent policies have lowered the priority of mathematical research not connected with the most immediate practical problems, even when that research would build up a store of knowledge for the solution of future practical problems. It appears, in fact, that little progress has been made in China toward developing new methods of OR.

5.2. PURE MATHEMATICS

Algebra

There is little research in algebra at this time in China. A small algebra group at the Institute of Mathematics in Peking works on the construction of algorithms. One algebraist, Wan Che-hsien, lectured to us in Peking. (See Abstract 6.2.4.3.)

At Peking University Tuan Hsüeh-fu conducts a seminar in group theory. It is effectively a course on basic material comparable to that found in some undergraduate courses in the United States. There are tentative plans to cover more advanced topics in the future. A few years ago Hung Chia-wei came out of such a seminar and wrote a nice paper in the theory of modular representations; he also obtained definitive results on a problem in optimal search. At Futan University Hsü Yung-hua works on general algebraic structures.

Topology

At the Institute of Mathematics there are 11 topologists; four of them now work in more applied areas of mathematics. Several people at Peking and Futan Universities have worked in topology, continue to be interested in it, but currently are not doing research in that subject.

Wu Wen-tsün and Wang Ch'i-ming have studied Dennis Sullivan's work on rational homotopy and have done further work in this area. (See Abstract 6.2.1.3.) Chang Su-ch'eng and a student of his have been computing cohomotopy groups using the spectral sequence obtained from filtering a CW complex by its skeletons. Several years ago some work was done in differential topology (e.g., immersions of manifolds in Euclidean space), but it was not continued because of the priorities that emerged from the Cultural Revolution. Members of the Institute's topology group have been studying *K*-theory and catastrophe theory.

The topologists are also searching for new applications of topology, but of course this is quite difficult. They work on topology per se when they are not at factories or involved in political discussion groups, an arrangement that seems to allow a fair amount of time for research. At present, the guess seems to be that no new people will be trained in topology for the next 5 years. The youngest topologist is 34.

In summary, some good work is being done in topology, but it is an area of low priority and there is considerable pressure on topologists to move into applied mathematics.

Analysis

A very small number of mathematicians are doing research in analysis. Some of the original work is really outstanding, and it is even more impressive when one takes into account the isolation in which it has taken place. In particular, work on analytic number theory and meromorphic functions is excellent. Recently there have appeared in *Acta Mathematica Sinica* a number of papers on partial differential equations and pseudo-differential operators; they show a mastery of the subject and constitute solid contributions. Homogeneous domains in several complex variables are also being studied. (See Abstract 6.3.3.2.) Some of the analysis done in conjunction with applied mathematics is described elsewhere in this report. It is our impression that there are virtually no young people being trained in analysis, and vast areas of analysis are not covered.

Analytic Number Theory

Excellent work in analytic number theory is done at the Institute by a group of Hua Lo-keng's students. The outstanding result achieved there in recent years is the theorem of Ch'en Ching-jun (1966; details published in 1973), which is the best result to date in the direction of the Goldbach conjecture. Recently Ch'en's proof has been simplified by other members of the Institute, Ting Hsia-hsi and Wang Yüan, as well as Pan Cheng-tung of Shantung University. (See Abstract 6.2.2.2.)

Although we found no one who worked in algebraic number theory, some joint work of Hua Lo-keng and Wang Yüan used properties of cyclotomic fields as well as deep results from analytic number theory in a problem in numerical analysis. (See Abstract 6.2.2.3.)

Complex Analysis

The most noteworthy contribution of Chinese mathematics in complex analysis lies in classical Nevanlinna theory, in work done by Yang Lo and Chang Kuang-hou of the Mathematics Institute of Peking. This area, which requires formidable analytic techniques, has been ploughed over carefully by many specialists all over the world for 50 years. Yang and Chang found something both new and deep to say about this venerable subject. (See Abstract 6.2.3.1.) To our knowledge, these two mathematicians are entirely isolated, although not so long ago there was a whole school of Nevanlinna theorists that grew out of the pioneering work of Hsiung Ch'ing-lai. There is interest in several complex variables in Peking, mostly as a result of Hua's work in symmetric bounded domains of the classical type. (See Abstract 6.2.3.2.)

Differential Geometry

There is some recent work on improving the Gauss-Bonnett theorem for manifolds with boundary (See Abstract 6.2.1.4.), and on the eigenvalues of the Laplacian on compact Riemannian manifolds. At Chekiang University there had been a school of classical geometers headed by Su Pu-ch'ing but it disappeared when its members became applied mathematicians. Some interest in the dynamical system aspects of foliations is evident at Peking University, but published results are not yet available.

6

ABSTRACTS OF LECTURES BY CHINESE MATHEMATICIANS

The delegation, during its visit, had the pleasure of hearing many lectures from Chinese mathematicians. This chapter presents abstracts of these lectures, in a style reminiscent of *Mathematical Reviews*.

6.1 APPLIED MATHEMATICS

6.1.1 Classical Applied Mathematics

6.1.1.1 Some Problems in the Approximate Solution of Operator Equations, Lin Chün, Institute of Mathematics

6.1.1.2 Mathematical Methods in Magnetic Exploration, Institute of Mathematics

6.1.1.3 A Study of the Global Solutions for Quasi-Linear Hyperbolic Systems of Conservation Laws

6.1.1.4 Analysis of Flow through a Turbo-Compressor Blade Passage, Wu Chung-hua, Institute of Mechanics

6.1.1.5 On Multiple Solutions of a Free Boundary Problem for the Equilibrium Equation of a Plasma, Chang Kung-ch'ing, Peking University

6.1.1.6 Recent Investigations in the Theory of Homogeneous Isotropic Turbulence, Chou P'ei-yüan and Huang Yung-nien, Peking University

6.1.1.7 The Propagation of Shock Waves in a Duct of Variable Cross Section and the Formation and Development of New Shocks, Li Hsüeh-tseng, Peking University

6.1.1.8 Dissipation of Tidal Energy in the Yellow Sea, Fang Kuo-hung, Institute of Oceanography

6.1.1.9 A Three-Dimensional Nonlinear Model of Tides, Feng Shih-tso, Shantung Oceanographic College

6.1.1.10 A Model for the Output of Retinal Ganglion Cells, Wang Yün-chiu, Institute of Biophysics

6.1.1.11 The Vibration of Shrouded Turbine Rotor Blades, Harbin Turbine Plant

6.1.1.12 Some Problems in the Design of the Shapes of Ships, Su Pu-ch'ing, Futan University

6.1.1.13 Boundary Value Problems for Partial Differential Equations of Mixed Type, Gu Ch'ao-hao, Futan University

6.1.1.14 The Scattering Operator in a Space of Indefinite Metric and a Singular Integral Operator, Hsia Tao-hsing, Futan University

6.1.1.15 Boundary Value Problem for a Self-Adjoint Elliptic Equation with Mixed Boundary Conditions, Li Ta-tsan, Futan University

6.1.1.16 Transmission of Pressure Step Signals in Pipelines, Liu Chao-jung, Futan University

6.1.1.17 Determination of the Dynamic Equation for the First Line Product of a Fractionating Tower, Li Hsün-ching, Futan University

6.1.1.18 Ray Tracing in Stratified Media, Yü Wen-chi, Futan University

6.1.1.19 Cam Design for a Loom, Wang Chien, Futan University

6.1.1.20 Flow Capacity of a Cable Wire Bundle, Shanghai Normal College

6.1.1.21 Motion Analysis of the Automatic Rice Planter, Ching Tang-mo and T'ang Yung-chi, Shanghai Normal College

6.1.1.22 Flow Analysis of Pipe Networks, Ts'ao Hsi-hua, Shanghai Normal College

6.1.1.23 Optimal Design of Piping Systems, Liu Chao-k'un, Shanghai Normal College

6.1.2 The Finite Element Method

6.1.2.1 The Finite Element Method, Feng K'ang, Institute of Mathematics

6.1.2.2 Infinite Similar Element Method for Calculating Stress Intensity Factors, Ying Lung-an and Kuo Chung-heng, Peking University

6.1.2.3 Stress Analysis of the Connecting Rod in Diesel Engine No. 6105Q, Wu Wei-ch'ang, Futan University

6.1.2.4 The Elastic Analysis of a Boiler End-Cap, Shanghai Normal College

6.1.3 Operations Research

6.1.3.1 Marshalling of Freight, Ma Chung-fan, Peking Hotel
6.1.3.2 Theory of Queues and Applications to Mining, Hsü Kuang-hui, Peking Hotel
6.1.3.3 The Use of Operations Research in the Operation of a Commune in Heilungkiang Province, Kuei Yüan, Harbin, in Hotel
6.1.3.4 The Use of Operations Research in Timber Production in Heilungkiang Province, Sung Pao-sheng, Harbin, in Hotel
6.1.3.5 Uses of OSM and CPM, Kuei Hsiang-yün, Institute of Mathematics
6.1.3.6 Cold Rolling of Steel, Yüan Te-tung, Shanghai Normal College

6.1.4 Statistics

6.1.4.1 Probability and Statistics, Institute of Mathematics
6.1.4.2 Scientific Research on Data Processing, W. T. Wu, Futan University
6.1.4.3 Application of Experimental Design, Ting Yu-en, Shanghai Normal College
6.1.4.4 Stochastic Integrals, Yen Chia-an, Peking Hotel

6.1.5 Industrial Applications

6.1.5.1 Diesel Locomotive Repair Plant in Peking
6.1.5.2 Visit to the Tach'ing Oil Field:
Separation of Oil and Water
Gas Chromatography
6.1.5.3 Visit to the Tongkiang Transistor Plant, Harbin
6.1.5.4 Feng T'ai Car Insulation Depot

6.2 PURE MATHEMATICS

6.2.1 Algebraic Topology and Geometry

6.2.1.1 The I^* Functor, Wu Wen-tsün, Institute of Mathematics
6.2.1.2 A Commutative Cochain Functor on Topological Spaces, Wang Ch'i-ming, Institute of Mathematics
6.2.1.3 Homotopy Numerical Invariants, Shen Hsin-yao, Institute of Mathematics
6.2.1.4 New Proof of the Gauss-Bonnett Theorem With Boundary, Yü Yen-lin, Institute of Mathematics

6.2.2 Number Theory

6.2.2.1 The Goldbach Hypothesis, Ch'en Ching-jun, Institute of Mathematics

6.2.2.2 On the Representation of Every Large Even Integer as a Sum of a Prime and an Almost Prime, Ting Hsia-hsi, Institute of Mathematics

6.2.2.3 Number Theoretic Methods in Analysis, Wang Yüan and Hua Lo-keng, Institute of Mathematics

6.2.3 Classical Analysis

6.2.3.1. Borel Directions and the Deficient Values of a Meromorphic Function, Yang Lo, Institute of Mathematics

6.2.3.2. Siegel Modular Domains, Hsü Yi-ch'ao, Institute of Mathematics

6.2.4 Algebra and Allied Fields

6.2.4.1 Non-Associative and Non-Distributive Rings, Hsü Yung-hua, Futan University

6.2.4.2 On the Problem of the Relationship between the Maximum and the Minimum Condition in Associative Rings, Hsü Yung-hua. Futan University

6.2.4.3 Binary Recurrent Sequences, Wan Che-hsien, Institute of Mathematics

6.2.4.4 Some Results in Operator Algebras, Li P'ing-jung, Institute of Mathematics

6.1 APPLIED MATHEMATICS

6.1.1 Classical Applied Mathematics

6.1.1.1 *Some Problems in the Approximate Solution of Operator Equations, Lin Chün, Institute of Mathematics*

A. <u>Convergence of iterative methods.</u>

1. Consider the problem $\Delta u = f(u)$ for a bounded domain Ω with $u = 0$ on $\partial\Omega$. The following two methods always converge locally to a solution:

Method I,

$$u_{n+1} = (1 - \tau)u_n + 2\tau v_n - \tau w_n,\ \Delta v_n = f(u_n)\ ,\ \Delta w_n = f(v_n),\ \tau > 0.$$

Method 2,

$$u_{n+1} = (1 - \tau)u_n + \tau v_n + \tau f'(u_n)w_n,\ \Delta v_n = f(u_n),\ \Delta w_n = u_n - v_n,\ \tau > 0.$$

2. Li, Chu, and Lin have proved two theorems, of which the following are special cases:

a. If $x = Bx$ has an isolated solution x^* and if $B'(x^*)$ has only a real spectrum, then $\exists\ \tau > 0$ such that $x_{n+1} = x_n - \tau(I - 2B + B^2)x_n$ converges locally to x^*. The convergence of Method I follows from this if the problem is written as $u = Bu$ with $B = \Delta^{-1}f$. For then $B'(u^*) = \Delta^{-1}f'(u^*)$ is completely continuous, so it suffices to show that its eigenvalues are real.

b. If $Ax = 0$ has a solution x^* with $A'(x)$ continuous at x^*, and if $\beta||y|| \geq ||A'(x_o)y|| \geq \alpha||y||$ with $\alpha > 0$, then $\exists\ \tau > 0$ such that $x_{n+1} = x_n - 2(\beta^2 + \alpha^2)^{-1}A'(x_o)Ax_n$ converges locally to x^*. The author hopes to apply these methods to the Navier-Stokes equations.

B. Maximum norm estimates for the finite element method.

1. The transport equation, being nonself-adjoint, does not come from a variational principle. Therefore, to apply the finite element method to it one can use least squares, following Bramble and Schatz. This yields equations that are self-adjoint, and positive definite, so they have a unique solution u_h. As basis functions in H^1, Lagrange interpolation polynomials of degree k are used. This leads to the inequality $||u - u_h||_{L^2} \leq \alpha h^k$. Then one uses the generalized Nitsche inequality $||P||_{C(T_h)} \leq \alpha h^{-1}||P||_{L^2(T_h)}$, where T_h is an element of diameter h, to get $||u - u_h||_{L\infty} \leq \alpha h^{k-1}$. (The simplest case of the transport equation is

$$\mu \frac{\partial u}{\partial x}(x,\mu) + u = \int_{-1}^{1} k(x,\mu,\mu')u(x,\mu')d\mu' + f(x,\mu)$$

on a rectangular domain.) This method has also been applied to study spatially periodic solutions of the Navier Stokes equations at small Reynolds numbers by Mr. and Mrs. Ding, Lo and Lin.

2. To treat the first boundary value problem for a fourth-order equation, such as the von Karman equation, the Galerkin method can be used with cubic splines. If the resulting equations have a unique solution, then one can show that

$$||u - u_h||_C \leq \alpha \left[||u - u_h||_{L^2} + ||u - u_h||_{H^2} \right]^{1/2} \leq \alpha h^{3/2}.$$

Next one uses the Markoff inequality $||P||_{C'(T_h)} \leq \alpha h^{-1}||P||_{C(T_h)}$ to get $||u - u_h||_{e'} \leq \alpha h^{1/2}$. This method can also be applied to the first boundary value problem for the Navier-Stokes equations.

C. Estimation and construction of eigenfunctions.

Let $\lambda_1 < \lambda_2 < \ldots$ and u_j be the eigenvalues and eigenfunctions of a self-adjoint linear operator A, and let v_j be approximations to the eigenfunctions, $j = 1,\ldots p$, with $(v_j,v_p) = \delta_{jp}$. Also let $||v||_1{}^2 = (Av,v)$. Then

$$\sum_{j<p} (\lambda_p - \lambda_j) \left|\left| u_j - \sum_{i<p} (v_i, u_j) v_i \right|\right|^2$$

$$\leq \sum_{j\leq p} \left[||v_j||_1{}^2 - \lambda_j \right] - \sum_{\substack{k>p \\ i<p}} (\lambda_k - \lambda_p)(v_i, u_k)^2.$$

From this it follows that

$$||v_p - (v_p, u_p) u_p||^2 \leq (\lambda_{p+1} - \lambda_p)^{-1}$$

$$\cdot \left[||v_p||_1{}^2 - \lambda_p + \frac{\lambda_{p+1} - \lambda_{p-1}}{\lambda_p - \lambda_{p-1}} \sum_{j<p} \left(||v_j||_1{}^2 - \lambda_j \right) \right].$$

As an example, suppose that there is a unique solution w of the equation $Aw = \tilde{\lambda} w + b$ where $(b, u_1) \neq 0$ and $\tilde{\lambda} < \lambda_1$. Let $\tilde{v} = \tilde{w}/||\tilde{w}||$ be an approximation to u_1. Then the above inequality yields

$$||\tilde{v} - (\tilde{v}, u_1) u_1||^2 \leq (\lambda_2 - \lambda_1)^{-1} [||\tilde{v}||_1{}^2 - \lambda_1]$$

$$\leq (\lambda_2 - \lambda_1)^{-1} (\lambda_1 - \tilde{\lambda}) \left[\frac{||b||}{|(b, u_1)|} - 1 \right].$$

Chu and Lin take $v^{(n)} = w^{(n)}/||w^{(n)}||$ as an improved approximation to u_1, where $Aw^{(n)} = w^{(n-1)}$, and show that $||v^{(n)}||_1{}^2 \lambda \rightarrow \lambda_1$.

The method can be extended to cover the transport equation and to yield convergence and error estimates for the discrete ordinate method, solving a problem posed by H. B. Keller in 1957.

6.1.1.2 *Mathematical Methods in Magnetic Exploration, Institute of Mathematics*

Two methods were described for finding the magnetic potential below the ground from observations on the surface. Laplace's equation was used, so the results are valid only in the homogeneous region above the ore body. One method is to use Fourier series and a separation of variables solution. It is an elementary calculation. Some other a priori information is needed to go further, since the problem is improperly posed. No applications of this solution have been made. The other method is to represent the solution as a sum of solutions due to specified ore bodies of given shapes and densities. Then the solution involves a set of unknown parameters. These are chosen to give a best (least squares) fit of observed data. A semi-empirical method of performing the search for the best fit was described.

6.1.1.3 *A Study of the Global Solutions for Quasi-Linear Hyperbolic Systems of Conservation Laws*

The results presented were those contained in a paper with the above title by Ting Hsia-hsi, Chang Tung, Wang Ching-hua, Hsiao Ling, and Li Tsai-chung, *Scientia Sinica* 16, No. 3 (1973), pp. 317-335. Existence in the large was proved for the initial value problem of one-dimensional gas dynamics, with more general initial data than had been considered previously. The results were also extended to a class of hyperbolic systems of conservation laws studied by Smoller and Johnson.

6.1.1.4 *Analysis of Flow through a Turbo-Compressor Blade Passage, Wu Chung-hua, Institute of Mechanics*

A detailed account of the form of the dynamical equations in a curvilinear, three-dimensional, non-orthogonal coordinate system was presented. The particular coordinate system to be used in a definite case uses as coordinate surfaces planes perpendicular to the machine axis, surfaces S_1 interpolated between the hub and the casing, and surfaces S_2 interpolated between adjacent blade surfaces. When the blade geometry is known, the S_1 surfaces are chosen in accord with experience as the best a priori guess for the stream surfaces of the flow. The quasi two-dimensional flow in each slice between adjacent S_1 surfaces is then calculated and its stream lines define the S_2 surfaces. The final flow calculation then treats the quasi two-dimensional flow between S_2 surfaces. When the blades are to be designed but surface pressures are given, the S_2 surfaces are chosen a priori and the procedure is that above with S_2 and S_1 interchanged.

It is claimed that this truncated procedure leads to results accurate enough for technological purposes; it certainly is much simpler and less demanding than a full three-dimensional analysis. It does seem clear, however, that major refinement would be needed for operating conditions that are significantly off-design.

6.1.1.5 *On Multiple Solutions of a Free Boundary Problem for the Equilibrium Equation of a Plasma, Chang Kung-ch'ing, Peking University*

Let $\Omega = \Omega_+ + \Omega_-$ be a given bounded plane domain with plasma in Ω_-, which surrounds Ω_+. Let $\psi_\pm(x)$ be the magnetic flux in $\Omega_\pm$. The problem is to find $\psi_\pm$ and $\Omega_\pm$ satisfying the following conditions:

$$L\psi_+(x) = -h[x,\psi_+(x),\lambda], \ x \text{ in } \Omega_+;$$

$$L\psi_-(x) = 0, \ x \text{ in } \Omega_-;$$

$$[\partial\psi/\partial n] = [\psi] = 0 \text{ on } \partial\Omega_+, \ \psi_-(x) = g(x) \text{ on } \partial\Omega.$$

Here $L = -\left[\partial_{x_1}{}^2 - \frac{1}{x_1}\partial_{x_1} + \partial_{x_2}{}^2\right]$, while g and h are given functions and λ is a parameter. This problem can be converted into a Hammerstein integral equation with a discontinuous kernel. Then, by using a result of Amann, *Journal of Functional Analysis* (1972), it can be proved that for all λ greater than a certain value λ^*, the equation has at least two positive solutions. This result is related to results of H. B. Keller, D. S. Cohen, T. Laetch, R. Temam, and P. Rabinowitz.

6.1.1.6 *Recent Investigations in the Theory of Homogeneous Isotropic Turbulence, Chou P'ei-yüan and Huang Yung-nien, Peking University*

The authors continue their work on homogeneous isotropic turbulence in *Scientia Sinica* (March-April 1975). There they include, as one of the characterizing parameters, the vortex Reynolds number Re. The new results include an extension to larger Re of the calculation of the decay of the correlation functions.

6.1.1.7 *The Propagation of Shock Waves in a Duct of Variable Cross Section and the Formation and Development of New Shocks, Li Hsüeh-tseng, Peking University*

Propagation of a shock wave through a duct of uniform cross section, continuing into one of increasing cross section, and then into a second duct of uniform cross section was studied numerically. Calculations were carried out in the following three ways: (a) inviscid quasi one-dimensional gas dynamics using the method of characteristics; (b) one-dimensional gas dynamics using a finite difference treatment with artificial viscosity; and (c) axially symmetric gas dynamics using finite differences with artificial viscosity dependent on the radial coordinate. Agreement was excellent among the results of the three methods. The third one provided a description of the mach stems that arise near the final junction. (See *Scientia Sinica*, November-December 1975.)

6.1.1.8 *Dissipation of Tidal Energy in the Yellow Sea, Fang Kuo-hung, Institute of Oceanography*

Jeffries has noted that 1/30 to 1/20 of the total dissipation of tidal energy in all oceans occurs in the Yellow Sea. However, equilibrium tidal theory yields the result $H_{s_2}/H_{m_2} = 1/2.15$ for the ratio of the height of the spring tide to the neap tide (component 2), whereas observation yields 1/2.73. Jeffries suggested that nonlinear friction might account for this discrepancy. The author constructed an approximate solution of a one-dimensional time-dependent tidal equation using a power law friction force $F = -k_p|u^{n-1}|u$ and applied it to data for the Yellow Sea.

6.1.1.9 *A Three-Dimensional Nonlinear Model of Tides, Feng Shih-tso, Shantung Oceanographic College*

A straightforward perturbation analysis was given leading to a system of linear problems. The equations are long and messy, and no solutions have been constructed.

6.1.1.10 *A Model for the Output of Retinal Ganglion Cells, Wang Yün-chiu, Institute of Biophysics*

Let a_i be the input to the ith rod or cone, $i = 1,\ldots,n$, and let g be the output of a particular ganglion cell. It is assumed that g is related to the a_i by

$$g = \sum_{r=1}^{m} w_r \left[\sum_{i=1}^{n} k_{ri} a_i \right]_b .$$

Here w_r and k_{ri} are given coefficients, with each k_{ri} equal to 0, +1, or -1, and $[f]_b = f$ if $f \geq b$, $[f]_b = 0$ if $f < b$ where b is given. For ganglion cells corresponding to *on* and *off* centers, respectively, the k_{ri} are given by

$$[k_{ri}]_{on} = \begin{bmatrix} -1 & +1 & +1 & +1 & -1 & 0 \\ 0 & -1 & +1 & +1 & +1 & -1 \\ 0 & 0 & -1 & +1 & +1 & +1 \end{bmatrix} = - [k_{ri}]_{off}.$$

As examples, consider the two inputs $[a_i] = (0,0,0,+1,0,0)$ and $[a_i] = (+1,0,0,+1,0,0)$, and choose $w_r = 1$, $r = 1,2,3$. Then for an *on* center, the two corresponding outputs are $g = +3$ and $g = +2$ respectively, while for an *off* center they are $g = -2$ and $g = -2$ respectively.

6.1.1.11 *The Vibration of Shrouded Turbine Rotor Blades, Harbin Turbine Plant*

An analysis of the free vibrations of uniformly spaced, identical-turbine blades on a cylindrical hub that are connected by identical and identically placed elastic bars. The characterization of the elastic interaction of adjacent segments of a blade was taken from a finite element analysis of a formulation in which there were five degrees of freedom (no radial displacement). A matrix equation was found for the eigenvectors whose 10 components describe the five (generalized) displacements and the five (generalized) forces. Graphs displaying the dependence on angular velocity of several of the eigenfrequencies were given for a particular design. The exciting frequencies appropriate to that design were also plotted against wheel speed. The significance of the results for operating policy was discussed; the general discussion included attention to the statistical question associated with non-identical blade configurations.

6.1.1.12 *Some Problems in the Design of the Shapes of Ships, Su Pu-ch'ing, Futan University*

A. One-dimensional problem.

1. In the case of small deflections, piecewise cubic splines are used, with y, y', and y'' continuous at the knots. There are n unknown quantities to be determined, namely $y''(x_i) - k_i$, $i = 1,2,\ldots,n$. There are only $n - 2$ coefficients, and it was decided to minimize

$$\sum_{2}^{n-1} |y'''(x_{i-0}) - y'''(x_{i+0})|^2,$$

instead of end point conditions.

Concerning the allowed locations for inflection points, it was stated that one cannot have two inflection points in adjacent intervals, and that there should be no inflection points in the first and last intervals. Furthermore, the first differences in curvature with respect to chord length should not have adjacent jumps in sign. The absolute value of the change of curvature at an inflection point should also be controlled.

2. In the case of large deflections, cubic parameter splines $x(t)$, $y(t)$ cubic in t in $t_i \leq t < t_{i+1}$ used with

$$t_i = \sum_{1}^{i} \overline{P_\nu P_{\nu-1}};$$

t cannot be arc length. A system of cubic basis functions and a variety of transformations related to the shape of the surface are used. In both (1) and (2), there was reference to the work of Ahlberg and Coons.

B. The faring of a surface.

Following the work of Carl de Boor (1962), one defines bicubic splines in a 2-dimensional mesh. They have the dimension $(n + 2)(m + 2)$ with $2(n + m) + 4$ boundary conditions. If they do not exist in practice, one uses faring conditions instead, which minimize sums of jumps in various partial derivatives in the two directions. The methods give satisfactory results in shipbuilding for interior surfaces on the side of a ship; they were used in particular on a 20,000-ton tanker.

6.1.1.13 *Boundary Value Problems for Partial Differential Equations of Mixed Type, Gu Ch'ao-hao, Futan University*

Friedrich's theory of symmetric systems of first-order partial-differential equations without regard to type was employed to prove existence and uniqueness for a large class of linear boundary value problems for equations of mixed type. The results are quite novel and of considerable interest. The problem was suggested by the difference between the formulations of two problems for equations of mixed type. They are the Tricomi problem, analyzed by C. Morawetz, and the problem of diffraction of pulses by wedges, treated by A. A. Blank and J. B. Keller. Some results on nonlinear equations were also mentioned.

6.1.1.14 *The Scattering Operator in a Space of Indefinite Metric and a Singular Integral Operator, Hsia Tao-hsing, Futan University*

The problem of finding the scattering operator for a quantum field theory in a space with an indefinite metric was formulated. It was converted into a singular integral equation that was solved by a series expansion.

6.1.1.15 *Boundary Value Problem for a Self-Adjoint Elliptic Equation with Mixed Boundary Conditions, Li Ta-tsan, Futan University*

The author considers a boundary value problem for a quasi-harmonic equation in a variable *u* in the case when there are different types of boundary conditions on different parts of the boundary--for example, the values of *u* are given on one part of the boundary, while on another part the values of its normal derivative are specified. Such problems arise in many practical cases--for example, in the torsion of a cylindrical bar with cylindrical holes. The problem can be solved by superposition of the standard solutions of several separate problems, one for each type of inhomogenous boundary condition, but this solution involves many complications, so that the resulting numerical accuracy is low. Instead, the problem can be solved by a suitable variational principle, where the quantity to be minimized involves one type of integral for each type of boundary condition. The finite element method can then be used to reduce this to a discrete approximation.

The author had discussed his methods with L. Nirenberg during the latter's recent visit to China.

6.1.1.16 *Transmission of Pressure Step Signals in Pipelines, Liu Chao-jung, Futan University*

In a manner equivalent to methods used in transmission line theory, the author used conservation of mass and momentum and the Laplace transform and adopted terminal impedances (and/or junction impedances) to provide a mathematical model for pulse transmission in a gas-containing pipeline. The resulting pulse was found to be the superposition of individual reflections, and its characterizing features were described.

6.1.1.17 *Determination of the Dynamic Equation for the First Line Product of a Fractionating Tower, Li Hsün-ching, Futan University*

The author adopted, without any mechanistic base, a linear ordinary differential equation with unknown coefficients to relate the influence of an input variable, such as tower temperature or flow rate, to the density of the fluid product. The coefficients were chosen to fit available data. Another equation was adopted to calculate the output with *several* input variables from the outputs given by single-input calculations. It was reported that the results provide a satisfactory description of the output.

6.1.1.18 *Ray Tracing in Stratified Media, Yü Wen-chi, Futan University*

Using geometric optics as a basic formulation and using triangular finite elements in the calculational scheme, the rays associated with the transmission and reflection of explosion-generated pulses over a complicated set of strata were computed. The times between successive events were calculated. Then an automated way to break plane regions up into triangular elements was described. The method is useful when many such subdivisions are needed in a technical problem.

6.1.1.19 *Cam Design for a Loom, Wang Chien, Futan University*

This was a student exercise using calculus to design a cam that produces a loom-harness motion meeting assigned constraints.

6.1.1.20 *Flow Capacity of a Cable Wire Bundle, Shanghai Normal College*

The problem of calculating the temperature distribution in a cable consisting of a bundle of insulated wires was considered. Joule heating in the wires is the heat source. The bundle was replaced by a single circular cylinder of uniform properties and then the problem was solved in an elementary way.

6.1.1.21 *Motion Analysis of the Automatic Rice Planter, Ching Tang-mo and T'ang Yung-chi, Shanghai Normal College*

The calculus was used in the analysis of the kinematics of the linkage used in the automatic rice planter. Some cumbersome trigonometry could have been avoided by the use of complex numbers; nevertheless, the analytic result permitted a comparison of peak acceleration and stress levels for various values of the parameters characterizing the motion.

6.1.1.22 *Flow Analysis of Pipe Networks, Ts'ao Hsi-hua, Shanghai Normal College*

A single-source pipe network is given, with the geometry of the network, impedances of the various branches, and flow demands at all sinks specified in advance. It is desired to find the flow value and pressure drop in each branch when the system is operating at capacity, i.e., when all flow demands are being met.

Let n_o denote the number of sinks, so that there are $n_o + \chi$ nodes in all, b the number of branches, and χ the number of circuits in a basis. Fix a set of fundamental circuits (a circuit basis), assign an orientation to each branch and each fundamental circuit, and let $\vec{A}$ (resp. B) denote the signed incidence matrix of nodes (resp. fundamental circuits) versus branches. For $1 \leq j \leq b$, let P_j (resp. Q_j) denote the signed pressure drop (resp. flow) in the jth branch, and for $1 \leq i \leq n_o$ let q_i denote the required net flow into the ith sink. With

$$P = \begin{pmatrix} P_i \\ P_b \end{pmatrix}, \quad Q = \begin{pmatrix} Q_1 \\ Q_b \end{pmatrix}, \text{ and } q = \begin{pmatrix} q_1 \\ q_{n_o} \end{pmatrix},$$

we have Kirchoff's laws:

1. $\vec{A}Q = q$ by conservation of mass
2. $BP = 0$ since the pressure drop around each circuit is 0.

Also for $1 \leq i \leq b$,

3. $P_i = s_i \mid Q_i \mid Q_i$,

where the value of the given impedance s_i depends on the diameter, length, and roughness of the ith branch. The problem is to solve the system (1), (2), and (3) of $n_o + \chi + b$ equations in the $2b$ variables $P_1, \ldots, P_b$, Q_1, $\ldots, Q_b$. Note that $n_o + \chi = b$ by a theorem of Euler on plane networks.

The nonlinear system (1), (2), and (3) is solved iteratively by a familiar linearization process that is equivalent to Newton's method. Let

$$S^{(o)} = \begin{pmatrix} s_1 & & & \\ & \cdot & & \\ & & \cdot & \\ & & & \cdot \\ & & & & s_b \end{pmatrix},$$

for $k \geq 1$ let $S^{(k)}$ be obtained from $S^{(k-1)}$ by solving the linear system

$$\vec{A}Q = q, \quad BP = 0, \quad P = S^{(k-1)}Q$$

to find $P^{(k)}$ and $Q^{(k)}$, and then set

$$S^{(k)} = \begin{pmatrix} s_1 \mid Q_1^{(k)} \mid & & & \\ & \cdot & & \\ & & \cdot & \\ & & & \cdot \\ & & & & s_b \mid Q_b^{(k)} \mid \end{pmatrix}.$$

From theorems on the convergence of Newton's method it follows that the sequences $P^{(1)}, P^{(2)},\ldots,$ and $Q^{(1)}, Q^{(2)},\ldots$ obtained in this way are convergent, say to P^*, Q^*. Then P^* and Q^* give the pressure drops and flow values that are of interest.

The above procedure has been applied successfully to existing networks with about 260 nodes and about 500 branches, and for such networks 10 to 20 iterations were required. The computation was terminated when $|Q_i^{(k)} - Q_i^{(k-1)}|$ was "sufficiently small" for all i.

It is anticipated that in the future the above sort of analysis will be applied to more complicated systems, such as those involving several sources, and also to pumping stations.

6.1.1.23 *Optimal Design of Piping Systems, Liu Chao-k'un, Shanghai Normal College*

It is desired to select the diameters of the pipes in a network of specified geometry and specified flow demands so as to minimize a cost function that involves both the diameters themselves and the minimum pressure P_N in the system. The contribution of the ith branch to the cost function is proportional to $(a + bD_i^{\alpha})\chi_i$, where D_i if the diameter of the pipe in question, χ_i is its length, and the parameters a, b, and α are independent of i. The parameter α ranges from 1.76 to 1.8, depending on the network in question. The D_i's were taken as continuous variables, even though in practice only a limited number of pipe sizes are available.

The presence of P_N was initially troublesome, but a change of variables made it possible to avoid confronting P_N directly and to cast the problem as one of minimizing a certain transformed nonlinear cost function subject to linear constraints. This problem was solved by SUMT (sequential unconstrained minimization technique), which involves the introduction of a penalty function for each constraint, construction of a new objective function that involves both the penalty functions and the original objective function, and the use of gradient methods to solve an unconstrained minimization problem for the new objective function. If the solution satisfies the original constraints, it is optimal for the original constrained problem; otherwise, the penalty functions are increased until a feasible solution for the original constrained problem is obtained.

On a sample problem the described method led to a network costing about 25 percent less than one produced by the old method of handling problems of this sort.

6.1.2 The Finite Element Method

6.1.2.1 *The Finite Element Method, Feng K'ang, Institute of Mathematics*

The finite element method goes back to Euler, but was long neglected. In 1943 Richard Courant introduced the modern concept of a finite element method, but his work did not attract much attention. About 1955 Clough and other structural engineers independently suggested the method, on empirical grounds. The mathematical contributions came later, beginning about 1966.

At the same time, one group in China working on these mathematical aspects published in the *Chinese Journal of Applied Mathematics and Numerical Mathematics* (Vol. 2, 1965, pp. 238-262) a paper summarizing three directions of this work:

1. A systematic method for solving elliptic boundary value problems, using a variational principle and simplicial approximations;
2. A general convergence theory for such problems;
3. Applications to elasticity problems, in terms of the principle of minimum energy.

Since the Cultural Revolution this work has been widely expanded. There is a library of programs for the applications of the finite element method to two- and three-dimensional cases, both static and dynamic. It includes elasticity problems for composite structures that fit naturally into a "finite element" decomposition. The method has been applied to the design of dams, gates of drydocks, dynamic response to seismic waves, etc.

In theoretical research the convergence problem for the so-called conforming case, in which an infinite-dimensional space is approximated by a finite-dimensional subspace, was settled by Irons and Strang. Feng studied the non-conforming case in which the function involved is not continuous. Here the appropriate finite-dimensional space is not a subspace, so the problem of convergence is more difficult. For a discontinuous but piecewise smooth function u on a domain Ω with boundary Γ, he starts with the Poincaré inequality,

$$\int_{\Omega} u^2 \, dx \, dy \leqq A \iint_{\Omega} (u_x{}^2 + u_y{}^2) \, dx \, dy + B \int_{\Gamma} u^2 \, ds,$$

where A and B are constants depending on Ω but not on u. The domain Ω is then triangulated. In the weakly discontinuous case in which u fails to be continuous only at isolated points, one-sided convergence is proved by suitable use of the Sobolev embedding theorem. In the strongly discontinuous case he uses a new energy norm, which is the usual Sobolev norm, plus a term depending on the discontinuities.

The finite element method has been applied to the analysis of thin shells. There are some convergence results, as yet incomplete.

Another important use is the elastic structure problem for a rectangular plate supported along an edge, with loading along the edge perpendicular to the plate. More generally, one may consider such problems for any complex; for example, heat conduction problems for a three-dimensional complex, with the three-dimensional Laplace equation on the three-dimensional cells, a two-dimensional heat conduction problem on the boundary 2-cells, etc. This leads to a suitable "graded" system of elliptic partial-differential equations, which can be analyzed in considerable detail.

6.1.2.2 *Infinite Similar Element Method for Calculating Stress Intensity Factors, Ying Lung-an and Kuo Chung-heng, Peking University*

For elastic solids with cracks the authors describe the displacement field in terms of two sets of elements. In the region near the crack tip, there are infinitely many elements, each geometrically similar to the other. The size is in proportion to the distance from the crack tip. This collection of similar elements makes up one major element of the several into which the whole object is subdivided. The displacements associated with the major elements are calculated in a conventional numerical way. The field in the finely subdivided region is obtained analytically by a long and intricate analysis that was not described. It will appear in *Scientia Sinica*.

6.1.2.3 *Stress Analysis of the Connecting Rod in Diesel Engine No. 6105Q, Wu Wei-ch'ang, Futan University*

Using a two-dimensional formulation, 701 triangular finite elements, and sensibly chosen approximate bearing load distributions, the elastic stresses in the connecting rod of a diesel engine in several of its operating configurations was calculated. The points of maximum stress were located and the results were said to be incorporated into current design procedures. Plans for more sophisticated treatments of the problem were described.

6.1.2.4 *The Elastic Analysis of a Boiler End-Cap, Shanghai Normal College*

This lecture was given by a third-year student. Using the principle of virtual work and triangular finite elements, the stresses in a cylindrical shell with a flat end were calculated for various radii of the fillet at the rim. It was determined that the stress levels were acceptable when the fillet radius was not less than two-thirds of the shell thickness and that the weld should be located not less than one shell thickness from the inside face of the flat end.

6.1.3 Operations Research

6.1.3.1 *Marshalling of Freight, Ma Chung-fan, Peking Hotel*

This lecture described work carried out by the Operations Research Division of the Institute of Mathematics in response to a request from the management of the Chinese railway system. In a freight classification yard, a train arrives with its cars in more or less random order, and before the train leaves the yard the cars must be recoupled in the order of destination. Thus, the cars with the nearest destination should be at the front of the train so they can simply be pulled off when that destination is reached, those with the second stop as destination should be next, etc. Recoupling is accomplished with the aid of *n* spur tracks, where usually $4 \leq n \leq 8$. The initial sequence of cars is decomposed into $\leq n$ subsequences by backing successive cars onto the various spurs, and the subsequences (with perhaps some cars at the end left behind in some pulls) can then be recombined in an arbitrary order as segments in a new sequence. There are various optimization criteria which might be used. In the problem that was solved, it is the following: For any initial sequence, the desired rearrangement should be accomplished with a minimum number of times that a collection of cars is *pulled* from one of the spurs. These are called "pulls"; the criterion takes no account of numbers of couplings, uncouplings, lengths of track, time, etc. For example, if 10 cars with possible destinations 1 through 7 are given in the order 6324135726, we wish to get them into the order 1223345667. On two tracks, this can be done by first backing

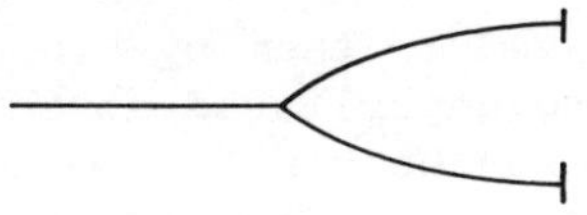

341536 onto the first track,
6237 onto the second track.

Pull both into the order 3415266237 (that's *two* pulls); then back them onto the two tracks in order

1223 onto the first track,
345667 onto the second track.

Then pull the first track's content onto the second, and pull out the whole train in the right order. Thus, it takes 4 pulls on 2 tracks to get the train together.

Define the *index* $k(\sigma)$ of the sequence σ = 6324135726 as follows: Start at the leftmost (in this case the only 1, put down all 1's, all 2's to the right of the last 1, 3's to the right of the last 2 if you have covered *all* the 2's, etc. In this case, the first subset defined in this way is 12 (positions 5 and 9). The next subset takes the other 2 and the second 3 (positions 3 and 6); it can't get to the first 3. The next subset takes the first 3, the 4, the 5, and the second 6; the last subset is 67. Thus 6324135726 has been decomposed into *4* non-decreasing, non-overlapping, non-descending sequences 12, 23, 3456, 67. The "4" is the index $k(\sigma)$ of the given sequence σ; the general definition is analogous; it is the number of times the ordering comes to the left end of the sequence.

Define n^{th} order Fibonacci numbers by

$$F_{-n}^{(n)} = F_{-n+1}^{(n)} = \ldots = F_{-1}^{(n)} = 0,$$

$$F_0^{(n)} = 1,$$

$$F_j^{(n)} = F_{j-1}^{(n)} + F_{j-2}^{(n)} + \ldots + F_{j-n+1}^{(n)}.$$

Theorem: *The minimum number of pulls in which a sequence* σ *can be ordered on* n *tracks is the integer* j *such that*

$$F_{j-1}^{(n)} < k(\sigma) \leq F_j^{(n)}.$$

Ma's results, and the role played in them by Fibonacci numbers and generalized Fibonacci numbers, are unpublished at this writing, but are reminiscent of results on "Polyphase Merging" reported on pp. 266-288 of Vol. 3 of Knuth's book on *The Art of Computer Programming*. The details given above were reconstructed from our notes with the aid of several combinatorialists; they are correct with high, but not unity, probability. According to Ma, the problem was mentioned without solution in some Swedish lecture notes on OR. He has not studied uniqueness of the solution, nor other optimization criteria such as numbers of kicks, numbers of uncouplings, or numbers of car movements. Questions under study include making up several trains at once, several yards side by side, and the advantages of some pre-classification.

6.1.3.2 *Theory of Queues and Applications to Mining, Hsü Kuang-hui, Peking Hotel*

First a *history* of the theory of the transient behavior of queues was given; references to Chinese work included M. I. Yüh (*Acta Mathematica Sinica*, 1959, pp. 494-502) who first did $M|M|n$ in a simpler way than Saaty (1960); and the speaker (*Acta Mathematica Sinica* 14, 1965, pp. 91-120) who did ξ_m, $\xi(t)$, W, and B (number of customers present at m^{th} arrival, number of customers present at t, waiting time, busy period) for $GI|M|n$ without the condition $\xi(0) = 0$ or 1. Others have been rediscovering his results in pieces ever since, i.e., N. U. Bhat in 1966 and 1968; J. DeSmit (thesis in 1971). The next topic was the structure of streams--general stochastic point processes. Necessary and sufficient conditions were given for a stream to be a Poisson process, or to be stationary (these are old Chinese results). Two new published pieces of work are: (a) Han Chi-yeh, treated queuing processes with arrival rate depending on queue length, did ξm, $\xi(t)$, and B for the case of one server; (b) work by the speaker on an *index* for how busy a service system is, with penalty for both idle system components and waiting customers.

Finally the application of queuing theory to mining was described. The problem is the lining-up of trucks to be unloaded after they arrive from different shovels. The variables are the number of "servers" (i.e., unloading facilities), the number of trucks, and the discipline of allocating trucks to shovels. (Centralized radio-controlled dispatching is not considered feasible.) Trucks come in different sizes. A big problem is: What to do when shovels break down? How do you allocate trucks to working shovels? They did a stochastic model, and a simulation, of this. (It is interesting that breakdown of shovels is an important variable.) There is much literature on transportation problems for open-pit mining, but no published direct attack on rescheduling.

6.1.3.3 *The Use of Operations Research in the Operation of a Commune in Heilungkiang Province, Kuei Yüan, Harbin, in Hotel*

The example presented emphasized the optimum scheduling of the labor force in a commune consisting of 8,000 acres, 10 production brigades, 34 production teams, and 11,000 population, including 2,500 agricultural workers. There is a separate flow chart for each season of the year, and the results were a saving of 8 percent of the people-days of work. A complete list of operations was made, and each operation was timed. The flow diagrams were then constructed, criticized, and revised. The major bottleneck (=principal contradiction) was alleviated by parallel and interwoven operations. For example, autumn operations include harvesting, selecting seeds for next year, transporting crops, irrigation, accumulating organic fertilizers, clearing fields, post-harvest plowing, threshing, delivering grain, field construction, tool maintenance, clearing the threshing ground, working on soil improvement, accounting, and record keeping. Each of these has a desired starting-time and a required number of worker-days; some have necessary predecessors or are prerequisites for others.

How did all this start? Hua Lo-keng visited the commune in 1974!

6.1.3.4 *The Use of Operations Research in Timber Production in Heilungkiang Province, Sung Pao-sheng, Harbin, in Hotel*

This talk contained a fascinating and impressive series of models used in all aspects of timber production, including logging, transportation by truck and rail, and operation of the sawmills. The work began as a result of five visits beginning in 1974 by Hua Lo-keng; since then, 401 "results" have been obtained. Some examples:

1. Operations analysis of the unloading of logs and the assignment of cranes to saws at the sawmill.
2. Statistical analysis of the time it takes trucks to make the trip

from the logging area to the sawmill. A very rough approximation to mean and variance of travel time was used to set a reasonable goal with the drivers--it should be possible for 90 percent of the trucks to make the trip in 450 minutes.

3. The sawmill is essentially in continuous operation. There are graphs of the number of trucks arriving per hour and the number unloaded per hour for each hour in a 24-hour period. However, one cannot operate equipment all the time. The solution to the critical path scheduling problem led to the borrowing and lending of workers among operatings.

4. An analysis of the details of the unloading operation. Of the two workers, one operates the winch and the other unloads. The sequence of steps includes hauling back the cable, engaging the main and secondary hooks, actual moving of the log, and disengaging the hooks. During the cable-hauling and the log-moving, the hook men were idle; the time was shortened by careful analysis of what could be done simultaneously.

5. Scheduling of the labor force between the construction of logging roads--grading, digging out stumps, etc.--and other activities. Also an optimal balance between labor and equipment.

6. Optimal loading of railroad cars at the logging site. Different piles of timber take different lengths of time to load, depending on their distance from the track. If the railroad schedule allows very little time, load the "easy" piles (close to the track); if there is plenty of time, load the tougher piles.

7. A very impressive overall schedule for locomotives in the loading, transportation, and unloading of timber. There are six locomotives to be cycled to eight different timber trains visiting different locations. Some locomotives are on a 2-day cycle, some on a 4-day cycle, in visiting different timber sites that are at different distances from the sawmill.

6.1.3.5 *Uses of OSM and CPM, Kuei Hsiang-yün, Institute of Mathematics*

A general description of the use, in the People's Republic of China, of the "optimal seeking method" (Fibonacci or golden section search for the optimum of an unimodal function of one variable) and the "critical path method" (use of CPM or PERT charts to discover production bottlenecks and, accordingly, to reallocate resources so as to speed up production), was presented. Much of the information supplied in this presentation, as well as in discussions with other members of the Operations Research Division of the Institute of Mathematics, is contained in the section on Operations Research and Optimization of this report.

6.1.3.6 *Cold Rolling of Steel, Yüan Te-tung, Shanghai Normal College*

In a four-stage rolling mill, electric motors of varying power drive each pair of rollers. The talk described a computational technique for adjusting the mill so that an equal percentage of the maximum power rating of each motor was utilized in the mill's operation.

6.1.4 Statistics

6.1.4.1 *Probability and Statistics, Institute of Mathematics*

This reports on two lectures given to summarize the work of the Institute of Mathematics on these topics.

They work on time-series, stochastic integrals, stochastic processes, experimental design, multivariate analysis, sampling theory, reliability theory, and non-parametric statistics. Applied problems mentioned include:

1. Breaking strength of fibers in a textile plant; the problem is to see if the mean breaking strength observed in different batches indicates that they came from the same distribution.
2. Manufacturing of glass insulators. Orthogonal experimental design allowed them to reduce the fraction of broken insulators from 0.01 to less than 0.001.
3. Standardization of clothing by cluster analysis.
4. The dependence of mechanical properties of steel on its chemical constituents.

Details on (2) and (4):

(2) The first part of the problem is the production of the glass, which must be annealed. In the orthogonal experimental design, they studied six factors and did 25 experiments with 16 replications each. Factors included temperature and speed of cooling; they were selected by the mathematicians and the workers together. Initially, when the workers did this themselves, they had too many factors.

(4) Multivariate analysis was applied to see what standards should be set for chemical composition and mechanical properties of a particular structured steel. They regressed the mechanical properties on the chemical components, looked at the residuals, found abnormalities, and tried to eliminate them. The abnormalities were explainable by the *heat* treatment of the steel, which was thus discovered to be a factor also. Polynomial transformations were used to improve the residuals. They stated that the problem had been reduced to that of the extreme value of a multiple integral and that multidimensional Mills' ratio was involved. They are still working on it. It came from a steel mill in Hopei Province and the results were reported at a conference of steel-makers.

6.1.4.2 *Scientific Research on Data Processing, W. T. Wu, Futan University*

The work of the group on digital processing of information was discussed in terms of the example of seismic signal processing. The problem is to find underground strata containing oil or natural gas. Data comes from recording of time-series caused by explosions; noise interference is the problem, and it is solved by better signal-to-noise ratios and filtering. They use the fast Fourier transform to do convolution and deconvolution, and they do not solve the integral equation as such to get velocity profiles. They claimed nothing new in this system, and in fact credited it to published work from M.I.T. and oil companies. They did claim that their computing was faster.

They also mentioned standard methods used to process electrocardiograms, satellite cloud pictures, and earthquake detection data.

6.1.4.3 *Application of Experimental Design, Ting Yu-en, Shanghai Normal College*

This talk on applications of experimental design covered a variety of examples of orthogonal experimental design. The examples included:

1. The improvement of gummed tape. This consists of synthesizing monomers into copolymer solutions and adding a cross-linking agent to the copolymer solution to form a gel. The experimental design involved nine factors, four interactions, and three levels of response.
2. Optimizing the flexibility and thickness of a rubber compound to be used between steel plates of a bridge.
3. Blue dye for a chemical factory.
4. Improving the output of a brewery.

The speaker noted that biology and chemistry students all learn experimental design. There is currently a movement to include it also in the last year of middle school.

During the question period, the speaker referred to a book to obtain further details on a problem. We thereupon requested a copy of the book, which we were given the next day. *Orthogonal Experimental Designs,* published by Shanghai Normal College in 1975, contains most of the examples from the talk. The table of contents of this book is reproduced in Appendix E.

6.1.4.4 *Stochastic Integrals, Yen Chia-an, Peking Hotel*

Yen reported on the work on random measures by Ch'en P'ei-te *(Acta Mathematica Sinica,* 1976 September), the work on innovation and filtering of random sequences by Tu and Pan, and his work on stochastic integrals including a simple and direct proof of a basic lemma by Paul André Meyer on the decomposition of local martingales.

6.1.5 Industrial Applications

6.1.5.1 *Diesel Locomotive Repair Plant in Peking*

This plant is charged with the periodic overhauling of diesel locomotives. The overhaul time, as measured by the number of plant working hours between the locomotive's arrival and completion of the overhaul, was formerly 120 hours. However, a careful PERT (or CPM) analysis of the process led to the discovery of several improvements, such as using more workers on parts of the task that were formerly most time-consuming or removing some of the parts from the main work line and renovating them elsewhere in the plant (done for parts of the task formerly involving very constricted work space). Details of the improvements, which yielded a reduction of the overhaul time to 40 hours, were explained by a worker with the aid of a PERT chart. Much was made of Chairman Mao's advice to "seek the principal contradiction," which could be translated in the present context as "look for the bottleneck."

6.1.5.2 *Visit to the Tach'ing Oil Field:*

Separation of Oil and Water When oil does not flow out of an oil well under the pressure of natural gas, it must be forced out. This is done by sending water down into the oil through a separate pipe near the well. As a consequence the oil is forced up out of the well, but it is mixed with water. The water is removed from the oil by first adding to it a chemical mixture called a demulsifier, then heating the oil, and finally separating the oil and water in a separation vessel. Experience suggests that the three most important factors controlling the efficacy of the separation process are the composition of the demulsifier, the temperature to which the oil is heated, and the setting of the values in the separation vessel. Therefore, each of these three factors was optimized by the optimal seeking method.

The demulsifier is a mixture of four ingredients. The quantities of the first, third, and fourth ingredients were fixed and that of the second was determined. Then the quantities of the first two were fixed and the amounts of the third and fourth were found by an alternating search. Next the optimal temperature was found and finally the optimal valve setting was determined. The criterion was the percent of water remaining in the oil, although the percent of oil in the wastewater was also considered.

Gas Chromatography The amount of water in the aluminum oxide in a refraction column was varied to yield the maximum separation ratio of two gases. The time required was also considered.

6.1.5.3 *Visit to the Tongkiang Transistor Plant, Harbin*

The plant has a total of 230 workers, 72 percent female, average age 26. They make more than 120 types of silicon devices and have had orders from more than 400 factories. Before the Cultural Revolution the plant had seven workers and repaired plastic raincoats. In 1974 Hua Lo-keng came and taught them optimal seeking methods. They have had more than 40 successes in applying them to a variety of production problems. Some samples are: optimum composition of chromium ion polishing solution, best temperature for welding, optimization of the cutting edge in an ultrasonic wave machine, optimizing pressure in a gasoline container that feeds a burner, optimum time for a sintering operation, and optimum time and temperature for photoetching. The plant contains a great deal of adapted old and scrap equipment.

The plant operates a July 21 "university" for the workers. Teaching is done by veteran workers and by invited guests from colleges and universities. The typical worker in the factory has graduated from middle school, although a fair number have been to primary school only. There are a pair of courses in mathematics and physics that contain material they specifically need and that last 1½ years at the rate of 8 hours per week. Some of this is during working hours and some is in the evening.

Workers' salaries depend on length of service but not on background. When more workers are needed a request is made to the state, typically once a year. It takes about 6 months to obtain a worker. Sometimes a specific technical background is requested; if the request is not met, training occurs in service.

In the beginning, the factory received a loan from the government, and further loans can be obtained when they are needed. It is a matter of pride to repay a loan in 1 year. The interest rate varies from 0.1 percent to 0.3 percent per year. Half the profit from the operation goes to the government; the other half is used partly for reinvestment and partly for "worker welfare"--e.g., improving school, day care, and medical facilities.

6.1.5.4 *Feng T'ai Car Insulation Depot*

The Feng T'ai Car Insulation Depot maintains and repairs refrigerator cars made in China and elsewhere (mainly East Germany before 1967). It has seven workshops and employs 1,520 workers. The depot can handle 12 cars simultaneously, and it overhauls 8-10 cars per month. Overhauls of varying degrees of thoroughness are scheduled every 6 months, every 1½ years, and every 6 years. We were told of three major problems for which mathematical methods have been employed; these were chosen from 372 "recent results." The plant started to use mathematical methods about a year before our visit. The interest began when they heard of successes elsewhere; they invited a teacher and 4 students from Peking Normal College to spend a month working with them, and they expect more students and faculty to come soon.

The three problems that they described to us were optimization of the compressor, of the insulation around pipes, and of the design of the oil burner in the car. Optimization of the compressor was done via a block design to examine five factors: suction pressure in the compressor, control of the position of the liquid level of ammonia, quantity of circulating salt water, concentration of salt water, and circulation of air in the motor. The insulation around the pipes used to be uric aldehyde, which was too soft, too water-absorbent, and not a good enough insulator. They now use polyurethane. A block design was used to find the optimal mixture of chemical, which turns out to be sensitive to external temperature and therefore different on different days. Optimal design of the oil burner concerns gravity flow of the oil into the combustion chamber; optimal search was used to optimize separately the size of the holes in the nozzle and the angle of the nozzle with the horizontal.

6.2 PURE MATHEMATICS

6.2.1 Algebraic Topology and Geometry

6.2.1.1 *The I* Functor, Wu Wen-tsün,* Institute of Mathematics*

Let K be the category of connected, simply connected, locally finite simplicial complexes and simplicial maps. A differential, graded, augmented algebra (DGA algebra) A with the derivative d is said to be minimal if it is free as a graded algebra and da is decomposable for all $a \in A$. Let M be the category of minimal DGA algebras over the real numbers.

In this lecture a functor $I^*: K \to M$ was defined as follows: For $X \in K$, let $A(X)$ denote the DGA algebra of differential forms (over the reals) as defined by Dennis Sullivan following ideas of Hassler Whitney. Roughly, an element of $A^p(X)$ is a real, polynomial differential p-form in the barycentric coordinates. If $X \in K$, $I^*(K)$ is defined to be a minimal subalgebra over $A(X)$. Sullivan proved that such a subalgebra exists and is unique up to an isomorphism.

Unlike homology or homotopy groups, the functor I^* on the constructions $X \times Y$, $X \vee Y$, SX and ΩX can be computed from $I^*(X)$ and $I^*(Y)$. Similarly, if $i: A \to X$, then $I^* \in V\ (XA)$ can be algebraically constructed from $I^*(X)$, $I^*(A)$ and $I^*(i)$, while if

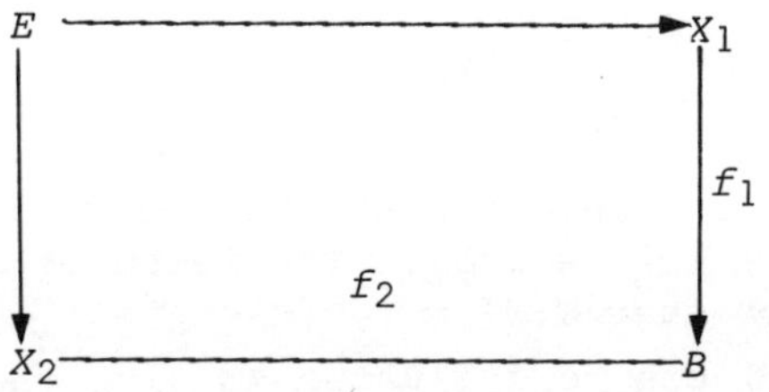

is a fibre square, then $I^*(E)$ can be constructed algebraically from $I^*(B)$, $I^*(X_i)$, and $I^*(f_i)$. These constructions were explicitly described for a fibre square and for attaching a cell to a space. This latter construction was utilized to develop an axiomatic characterization of I^*.

Finally, I^* on Lie groups and homogeneous spaces was discussed; it was shown that if G is compact Lie group and H is a closed subgroup, then $I^*(G/H) \approx H^*(G/H)$.

6.2.1.2 *A Commutative Cochain Functor on Topological Spaces, Wang Ch'i-ming, Institute of Mathematics*

For a simplicial complex K, let $A(K)$ denote the DGA algebra of rational differential forms as defined by Sullivan. This talk described an extension of A to semi-simplicial sets and hence to topological spaces.

*See also Wu Wen-tsün, the I^* Functor, *Acta Mathematica Sinica* 8(4): 162-172.

$A(K)$ and $C^*(X;Q)$ were shown to be chain-equivalent using Andre Weil's method for proving de Rahm's theorem. A theorem of G. Hirsch on the cohomology of fibrations and a theorem of Wu Wen-tsün on minimal DGA algebras for a fibre square are extended to the case of non-compact fibres whose cohomology is of finite type.

6.2.1.3 *Homotopy Numerical Invariants, Shen Hsin-yao, Institute of Mathematics*

This talk described methods of computing the cohomotopy groups, $\Sigma^n(X)$ of a *CW*-complex X. The spectral sequence obtained from the relative cohomotopy groups $\Sigma^n(X^m, X^{m-1})$ and primary- and higher-order cohomology operations were utilized in these studies.

Suppose X is a finite *CW* complex of dimension $n + 2$. A collection of numerical invariants were associated to X by considering the dimensions of the kernel and cokernel of Sq^2 on $H^n(X)$, the secondary operation associated to Sq^2Sq^2, and related operations involving the Bockstein operations. As an abstract group $\Sigma^n(X)$ was then computed from these numerical invariants.

6.2.1.4 *New Proof of the Gauss-Bonnett Theorem with Boundary, Yü Yen-lin, Institute of Mathematics*

Let M be a Riemannian polyhedron, i.e., a compact oriented Riemannian manifold with boundary, where the boundary is a finite union of smooth manifolds. Let $E(\Omega)$ be the Gauss-Bonnett integrand. If $p:S(M) \to M$ is the unit sphere bundle, then Chern's work shows that $-p^*E(\Omega) = d\pi_{ch}$ for some differential form π_{ch} in $S(M)$. The lecturer shows that there exists a chain A in $S(M)$ such that $X(M) = \int_M E(M) + \int_A \pi_{ch}$. Furthermore, he shows that the last integral can be evaluated to give the sum in the Allendoerfer-Weil version of this theorem. This accomplishes the task of giving an intrinsic proof for Riemannian polyhedra of the Allendoerfer-Weil formula, which was proved by them using unpleasant extrinsic methods.

The essence of the argument lies in carefully defining the chain A used for the second integral. Roughly, it goes as follows: If $x \in \partial M$ and x is a smooth point of ∂M, define A_x to be the unit outer normal of ∂M at x. If x is a corner, let

$$\{A_x = X \in M_x \mid |X| = 1 \text{ and } < X, Y > \leq 0 \text{ for all inward } Y\}$$

$$= \text{"dual cone of } \partial M \text{ at } x."$$

Then the chain A is taken to be the union of the A_x, for all $x \in \partial M$. One shows that if η is a smooth unit vector field on M with an isolated zero y in the interior of M, then $\eta: M \to SM$ except at y and $\eta(\partial M) = A + \partial c$ for some chain c in SM. From this formula, the above assertions follow readily.

There should be other nontrivial applications of the preceding formula for chains, but none were mentioned in the lecture.

6.2.2 Number Theory

6.2.2.1 *The Goldbach Hypothesis, Ch'en Ching-jun, Institute of Mathematics*

The author reported on his remarkable result that every sufficiently large even integer is of the form $p + c$ where $c = p_1$ or $p_1 p_2$ with p and p_i primes. This is a result of major importance in additive number theory. In proving this he introduced a new idea and was able to estimate

$$\Omega(x) = \sum_{\substack{p_3 \leq x/p_1 p_2 \\ x-p = p_1 p_2 p_3 \\ x^{1/10} < p_1 \leq x^{1/3} \leq p_2 \leq (x/p_1)^{1/2} \\ p, p_i \text{ primes.}}} 1$$

Details were not presented, but are contained in his paper, "On the Representation of a Large Even Integer as the Sum of a Prime and the Product of at Most 2 Primes," *Scientia Sinica* 16 (1973), pp. 157-176. The result was first announced in *Kexue Tungbao*, 17 (1966), in Chinese (MR *34* [1967] #7483).

He also reported on the following result about the distribution of almost primes in an interval.

Let α be the smallest number such that there exist 2 integers n with $x - x^{\alpha} < n \leq x$, where each n is a product of at most 2 primes for all sufficiently large x. Then $\alpha \leq 1/2$.

Apparently the conjectured value is $\alpha = 1/2$.

6.2.2.2 *On the Representation of Every Large Even Integer as a Sum of a Prime and an Almost Prime, Ting Hsia-hsi, Institute of Mathematics*

Ting Hsia-hsi reported on joint work with Pan Cheng-tung and Wang Yüan that simplified Chen's estimate for Ω and yielded a simpler proof of

Chen's first result above. (*Scientia Sinica* 18 [1975], pp. 254-282 and 599-610)

6.2.2.3 *Number Theoretic Methods in Analysis, Wang Yüan and Hua Lo-keng, Institute of Mathematics*

References: "On Uniform Distribution and Numerical Analysis I, II, III," *Scientia Sinica* 16 (1973), pp. 483-505; 17 (1974), pp. 331-348; 18 (1975), pp. 184-198.

The problem is to approximate $\int_G f$ by a finite sum, where G is the unit cube in n dimensions and f satisfies some restrictions. This method is also used to get a numerical solution to Volterra's integral equation

$$\phi(x) = \int_0^x K(x,y)\phi(y)dy + f(x)$$

The method is to choose a set of points in the unit cube that are uniformly distributed and can be constructed in a simple manner. The basic idea for choosing such points is to use properties of the ring of integers in the real cyclotomic field $Q(\zeta + \bar{\zeta})$, $\zeta = e^{2\pi i/p}$, p a prime. Error estimates are computed but the constant in the error cannot be explicitly bounded since it depends on the noneffective results of W. Schmidt. A key point is to choose a unit in $Q(\zeta + \bar{\zeta})$, which is a *PV* number. The methods are based on deep number-theoretic results. Although the constant in the error term cannot be determined, the authors deal with some numerical examples that indicate that the method should yield good approximations.

6.2.3 Classical Analysis

6.2.3.1 *Borel Directions and the Deficient Values of a Meromorphic Function, Yang Lo, Institute of Mathematics*

This is a report on the joint work of Yang Lo and Chang Kuang-hou in the past 3 or 4 years, with some theorems yet to be published. This represents an outstanding contribution to Nevanlinna theory. The results are deep and original.

Let f: $\mathbf{C} \rightarrow \mathbf{C} \cup \{\infty\}$ be a meromorphic function of order ρ. Valiron proved long ago that if $0 < \rho < \infty$, then there exists a ray A emanating from the origin such that if $n(r,\varepsilon,\alpha)$ denotes the number of zeros of $(f - \alpha)$ (counting multiplicity) in the set $\{z: |z| \leq r$, and $|\text{angle } A - \text{angle } z| \leq \varepsilon\}$, then with the exception of at most two α's,

$$\limsup_{r \to +\infty} \frac{\log n(r,\varepsilon,\alpha)}{\log r} = \rho,$$

whenever $\varepsilon > 0$. Such a ray is called a *Borel direction* of f. The existence of a Borel direction intuitively means that the meromorphic function must exhibit quite regular behavior along some radial direction. The following are the theorems of Yang and Chang.

Theorem 1. *Let* $0 < \rho < \infty$ *and let* p *be the number of deficient values of* f *(i.e., of those* $a \in S^2$ *such that* $\delta(a) > 0$*, where* δ *is the deficiency function of Nevanlinna theory). If* q *is the number of Borel directions, then*

$$p \leq q$$

and the inequality is optimal.

Theorem 2. *If* $0 < \rho < \infty$*,* f *is entire,* p *denotes the number of finite deficient values, and* q *denotes the number of Borel directions, then* $p \leq q/2$*; this inequality is optimal. If* $q < \infty$*, then* $p < 2\rho$.

Theorem 3. *With notation as in Theorem 1, let* P_e *denote the number of finite nonzero deficient values of* $f^{(e)}$ *(the* e^{th} *derivative of f). Then* $\Sigma_e P_e \leq q/2$ *and* $q < \infty \Rightarrow \Sigma_e P_e < 2\rho$.

The following are theorems of Yang and Chang pertaining to the distribution of Borel directions. In general, the distribution is completely arbitrary. Indeed, if *E* is the set of angles on the unit circle that correspond to the Borel directions of a meromorphic *f*, then *E* is closed in the unit circle. Conversely,

Theorem 4. *Given* ρ*,* $0 < \rho < \infty$*, and a nonempty closed set on the unit circle, there exists a meromorphic function* f *of order* ρ *with* E *as its set of Borel directions.*

Theorem 5. *Let* f *be a meromorphic function of order* ρ*,* $0 < \rho < \infty$ *and let* f *have at least one deficient value. If* f *has a* q *Borel direction, then either* $q = 1$ *and* $\rho < 1/2$*, or* $q > 1$ *and there exist two Borel directions separated by an angle* $\leq \pi/\rho$.

If *f* is actually entire, the preceding theorem can be sharpened.

6.2.3.2 *Siegel Modular Domains, Hsü Yi-ch'ao, Institute of Mathematics*

This is a continuation of earlier work by the author as, for example, in his paper, "On the classification of symmetric schlicht domains in several complex variables," Shuzue Jinzhan (1965), pp. 109-144 (Chinese) and *Mathematical Reviews* 33 (1967), #5936. His work builds on the work of Pjatetskii-Shapiro ("Automorphic Functions and the Geometry of Classical Domains; English translation published by Gordon and Breach, New York, London, and Paris, 1969). He proved that every homogeneous bounded doman *D* can be realized in the form of a Siegel domain of genus 1 or 2; the simplest example of a Siegel domain of genus 2 is the set of all pairs of complex numbers *z*, *u* with $Im\ z - |u|^2 > 0$.

In 1959-1961 Pjatetskii-Shapiro introduced the Siegel domains of genus 2

$$D(V,F) = \left\{(z,u) \in \mathbf{C}^n \times \mathbf{C}^m \middle| Im\ z - F(u,u) \in V\right\}$$

Here V is a convex cone in $\mathbf{R}^n$, not containing a whole line, and $F: \mathbf{C}^m x \mathbf{C}^m\ \overline{V}$ is hermitian, linear in its first argument, with $F(u,u) = 0$ if and only if $u = 0$. The author now defines $D(S_N)$ a *normal* homogeneous Siegel domain, in terms of a suitable family S_N of matrices of complex numbers. He proves:

Theorem 1. *Any homogeneous bounded domain is isomorphic to a normal homogeneous Siegel domain, and* $D(S_N)$ *is isomorphic to* $D(\tilde{S}_N)$ *if and only if* S_N *and* $\tilde{S}_N$ *are suitably* σ *equivalent, where* σ *is a permutation of* $N + 1$ *letters.*

Theorem 2. *Any* $D(S_N)$ *of "square type" is indecomposable.*

Theorem 3. *The* $D(S_N)$ *of the "first kind" can be arranged in five classes and those of the "second kind" in three classes.*

6.2.4 Algebra and Allied Fields

6.2.4.1 *Non-Associative and Non-Distributive Rings, Hsü Yung-hua, Futan University*

The "standard algebras" of A. A. Albert (1948) and R. D. Schafer (1968) are further extended to NAD rings R (non-associative and non-distributive rings) defined as follows. R is an additive abelian group; there is a set F of scalars that act as left *and* right distributive operators on R; there is a binary operation of multiplication (but no identities are required). Then consider subsets A or R closed under the scalar operators. The multiplication of R now assigns to $b \in R$ and A a set $[A,b]$ of all products ab for $a \in A$ and a set $[b,A]$, both sets being closed under the scalars F. This set-valued (multivalued) product then is required to satisfy analogs of the associative and distributive laws. The decomposition of a semi-simple NAD ring into such simple rings is then established, while ideals, hyper-nilpotent ideals, the radical, and related concepts are then extended to such rings. There are specific examples of such NAD rings arising from transfinite power series rings.

6.2.4.2 *On the Problem of the Relationship between the Maximum and the Minimum Condition in Associative Rings, Hsü Yung-hua, Futan University*

In 1939 Hopkins proved that if an associative ring has a unit, the minimum condition implies the maximum condition. Later Y. Akizuki and I. S. Cohen proved that if a commutative and associative ring R has a unit, then the minimal condition for ideals is equivalent to a pair of conditions (1) and (2), as follows:

1. The maximal condition for ideals
2. Every prime ideal is maximal

The author generalizes this theorem. Dropping the condition that R has a unit, he proves that the stated equivalence holds if and only if there exist positive integers k_i such that $k_i N^i$ is contained in RN^i, for $i = 1, \ldots, m$. Here N is the nilradical of R and m is the least integer with $N^m = 0$ and $N^{m-1} \neq 0$.

Using the same condition on N, he also generalizes Hopkins' Theorem.

6.2.4.3 *Binary Recurrent Sequences, Wan Che-hsien, Institute of Mathematics*

Let F be the field of two elements. If $f(x_1,\ldots,x_n) \in F[x_1,\ldots,x_n]$, a binary recurrent sequence is defined by initial values $a_o,\ldots,a_{n-1}$ and for $k \geq 0$ by the equation

$$a_{n+k} = f(a_k,\ldots,a_{k+n-1}).$$

The author discussed some general properties of such sequences.

If $f(x_1,\ldots,x_n)$ is linear and homogeneous, the sequence is called a linear recurrent sequence. Thus

$$x_{n+k} = c_1 x_k + c_2 x_{K+1} + \ldots + c_n x_{k+n-1}, \; c_i \in F.$$

It may be assumed that $c_n \neq 0$; otherwise one could do with $n - 1$ variables. Thus $c_n = 1$. The characteristic polynomial of the sequence is defined by

$$G(x) = 1 + c_1 x + \ldots + c_n x^n.$$

A linear recurrent sequence is clearly periodic of period at most $2^n - 1$. If the period is $2^n - 1$, then the roots of $G(x)$ *are primitive* $(2^n - 1)^{th}$ roots of 1. In this case the sequence is called an n-sequence.

Let α be a primitive $(2^n - 1)^{th}$ root of 1. Then some criteria are found to give an algorithm that allows one to compute explicitly the minimum polynomial of certain powers α^t of α with $(t, 2^n - 1) = 1$.

6.2.4.4 *Some Results in Operator Algebras, Li Pi'ng-jung, Institute of Mathematics*

I. *Real C*-algebras* In 1943 Gelfand and Neumark proved that a complex B^*-algebra S that satisfied

$$\text{(i)} \quad ||x^*x|| = ||x^*|| \; ||x||$$

and;

$$\text{(ii)} \quad ||x^*|| = ||x||$$

for all $x \in X$ is isometric and isomorphic to a uniformly closed operator *-algebra in some complex Hilbert space. In 1960, Glimm and Kadison ("Unitary operators in C^*-Algebras," *Pacific Journal of Mathematics* 10 [1960], pp. 547-556; *Mathematical Reviews* 22 [1961], #5900) proved that condition (ii) could be omitted in this theorem, i.e., the * operation in B^* algebras is automatically isometric. This lecture established for the first time a corresponding result for real B^*-algebras that are hermitian (every h with $h^* = h$ has a real spectrum): Any real hermitian B^*-algebra with an identity that satisfies the condition

$$x^*x = xx^* \text{ implies } ||x^* x|| = ||x^*|| \cdot ||x||$$

is a real C^*-algebra. This result is published in *Acta Mathematica Sinica* 18 (1975); English translation by the machine translation project, The Chinese University of Hong Kong.

II. *Tensor products of C^*-algebras* Let A⊗B be the algebraic tensor product of two C^*-algebras A and B. *A norm* α on A⊙B is called a C^*-norm if

$$(\alpha(u))^* = \alpha(u^*)$$

for all u in A⊙B, and is called a cross-norm if $\alpha(a \odot b) = ||a|| \ ||b||$. In 1956 Takesake ("A Note on the Cross-norm of the Direct Product of Operator Algebras," *The Kodai Mathematical Seminar Report* 10 [1958], pp. 137-140; *Mathematical Reviews* 20 [1959], #6666) used the faithful representation of A and B to construct an explicit C^* norm α_o that is also a cross-norm. In 1972, S. Sakai proved that if A or B has an identity, this α_o is the smallest C^*-norm in A⊙B, and every C^*-norm on A⊙B is necessarily a cross-norm. Now the role of the identity can be clarified as follows, letting $A + 1$ denote the algebra A with an identity adjoined: Any C^*-norm on A⊙B has a unique extension to $(A + 1)\odot(B + 1)$, and

$$(A + 1) \otimes_\alpha (B + 1) \cong A \otimes_\alpha B + (A + 1) + (B + 1),$$

where $\otimes_\alpha$ denotes the completion under the norm α of the corresponding space with ⊙.

III. *Spectral resolutions of commuting self-adjoint operators* Let A be a self-adjoint operator on a Hilbert space H. A theorem of Stone (1932) asserts that when A has a cyclic vector, there is an isomorphism of H to an L^2-space for a suitable finite Borel measure ν on $\mathbf{R}$ so that A becomes multiplication by some λ. The corresponding question for a countable family A_i of commuting self-adjoint operators has been considered by J. M. Jauch and B. Misra ("The spectral representation," *Helvetia Physica Acta* 38 [1965], pp. 30-52; *Mathematical Reviews* 35 [1968], #773) for the special case that the measure ν is absolutely continuous. The author now establishes a more general version of this theorem.

7

MATHEMATICAL EDUCATION

7.1 IN THE UNIVERSITIES

Of all the facets we saw of Chinese society, education seems to have been the most profoundly affected by the Cultural Revolution. From the anti-elitist, anti-bourgeois campaigns emerged a new educational system based on the philosophy that "education must serve proletarian politics and must be combined with productive labor." According to a spokesman from Peking University, "The task of the university involves both the study of books and political and ideological work. The educational system in the United States is constructed to teach students to believe in the capitalist system and that the capitalist system is eternal. Here in China we teach the students never to take the capitalist road. In science and in culture we strive to catch up to the advanced level in other countries. We want students to be both red and expert. We do not wish to shut up students in the campus. Hence we follow the open-door way for the university."

We were repeatedly told that the current program in mathematical education is experimental. However, the basic principles of the program seem to be fixed. Two dominant guidelines of the new system are (a) "combine theory with practice" and (b) "place politics in command." The interpretation and application of these frequently cited slogans are discussed below in the context of mathematical education.

"Combine Theory with Practice"

To achieve the combination of theory with practice in mathematical education, the students and teachers engage in "productive labor" in a factory as an integral part of any mathematics course. For example, the course in first-year calculus in Peking University last year was illustrated throughout with the calculation of where to support a horizontal laser tube to minimize the maximum deflection. Similarly, at the Shanghai Normal University, the first-year calculus course was centered on the design of a cam for a weaving machine. The topics taught in these courses were mostly motivated by these applications. In both cases we were told that the students and teachers discovered these problems through their experiences in factories, and that later, in the classroom, they learned enough mathematics (including some numerical analysis) to provide practical answers to these problems. Our discussions at Tsinghua University, Futan University, and the May 7th Middle School in Shanghai, as well as discussions with teachers in Harbin, indicated that this method of teaching is common to all universities in China today.

The following table shows the 36-month schedule of study for students specializing in mathematics at Futan University. This table was constructed for us after much discussion and consultation with students, and our impression was that it represented a kind of possible scheme rather than a fixed program of study. The program described has a total enrollment of 120 students. The mathematics department also has a program in mechanics that enrolls an additional 136 students.

First period (8 months)

2 months	3 weeks	4 months	1 month
Calculus (6 hours/week*) Algorithms (4) Physics (4)	Work in factory	Calculus and solid geometry (6) Curves and surfaces (4) General physics (4) Analyze and solve problems in the above classes	Analyze and solve problems Independent projects in small groups and self-study

Second period (8 months)

1 month	5 months	2 months
Work in factory and study mechanics there	Finite element method including linear algebra (6) Calculus (4) Physics (2)	Solves practical problems (sometimes in factory)

Third period (10 months)

4 months (including at least 1 month in factory)	6 months (including at least 1 month in factory)
Mathematical methods of industrial process control--one variable	Mathematical methods of industrial process control--several variables Probability and statistics (lectures) for 6 weeks, 2 classes per week

Additional time as follows:
1 month in military work
1 month in harvest work (total in 3 years)
4 months on graduation project
2 weeks/year winter vacation
4 weeks/year summer vacation

TOTAL = 36 months

*Hours/week is implied for the rest of this table.

"Place Politics in Command"

"Proletarian politics" refers to the class struggle in general and, more particularly at the current time, to the struggle between the "correct line" of Mao Tse-tung and the "revisionist line." It is said that the revisionists de-emphasize class struggle in favor of the promotion of stability and unity and the development of the national economy. They disagree with the post-Cultural Revolution educational system, claiming that the shortened and simplified curriculum, the requirement for all middle-school graduates to work in a commune or factory before admission to university, and the enrollment of a large proportion of workers, peasants, and soldiers in the universities are delaying the fulfillment of China's professional scientific and technical needs. These attitudes are considered by their critics to encourage bourgeois thinking and class distinctions.

At the time of our visit, the opposite view was in favor among those responsible for educational policy. The stress on proletarian politics is aimed at eliminating any tendencies of students to feel superior to manual laborers. Through the open-door method, whereby students work side by side with workers and peasants, students are taught to identify with and learn from the workers and peasants. Each week students participate in several political discussion sessions, either at the university or at a factory or commune, to exchange ideas about the class struggle and to "grasp the correct line." Students and workers are expected to take part jointly in political struggles.

"Proletarian politics" is an essential ingredient in considering students for a university. The system was described to us at Peking University in the following way:

> Education must serve proletarian politics and must be combined with productive labor. Students must take an active part in revolution and construction. Students at the university come from workers, peasants, or soldiers who have finished middle school.
>
> The primary schooling extends for 5 years and is followed by 5 years of middle school. The students then go to the countryside or factory. After 2 years of this practical experience they can then apply for admission to university or college.
>
> The local educational authorities then form a subcommittee to examine the various applications and again select. The cases selected are then submitted to the university for review. In this review, the ideological and political qualifications come first. The vocational level and practical experience is also considered.

Very few students are accepted, and the student-teacher ratio is very low, as illustrated by the following table. (See Table 1.) There has been almost no graduate education in mathematics, although there are some tentative graduate programs scheduled to begin next year.

TABLE 1 Number of Students and Faculty at Five Universities

Institution	Number of Full-Time Students	Number of Faculty	Number of Students in the Math Dept.	Number of Faculty in the Math Dept.	Number of Correspondence Course and Short-Term Students
Peking University	6,000	2,700	500	N.A.	50,000
Tsinghua University	10,000	3,000[a]	NO MATH DEPT.		20,000
Heilungkiang University	2,000	1,000	210	90	--
Futan University	3,000	4,000 (including staff workers)	256	111 teachers, 33 lab technicians	10,000
Shanghai Normal University	7,000	2,000	755	185	15,000

[a]9,000 working staff including personnel and workers.

Inherent in any program of education are certain assumptions about human nature. It seems to us that one of the basic differences in assumptions between the current Chinese and the Western educational system is the notion of talent and genius. In accordance with the "correct line," it is considered that talent has only a minor effect on a person's capacity, and that genius does not exist. True dedication to serve the people and the needs of the state is believed adequate to overcome any differences there might be in talent. Current policy does not condone the idea that a person may be inspired by the inner beauty of science or mathematics. (See Appendix D-3.)

It seems to us that there are two main goals of the Chinese educational system. The first and currently paramount concern is to promote egalitarianism and a socialist consciousness, preventing a sense of elitism among the educated. The second is to provide the training required to build up modern technology in China. It is very hard for us to judge the extent to which these goals are being met.

In relation to the first goal, we often asked students about their backgrounds. Several described themselves as peasants, but upon further questioning, they clarified by saying that they were born in the city but had become peasants by spending from 2 to 5 years on a commune or state farm, where they had worked and studied politics. Similarly, some students who were classified as workers seemed to have obtained their working-class background after graduation from middle school. As a result, the categories of workers, peasants, and soldiers have become very broad.

Judging the effectiveness of the system in fulfilling the second goal is almost impossible. The new courses are highly experimental and there seems to be no method to evaluate their effectiveness. Although the students are tested on their political attitudes and knowledge, it is claimed that there is no comparable evaluation of their technical proficiency.

One may ask whether the current educational system will produce enough people sufficiently trained in mathematics to meet the needs of modern technology, which China is in the process of developing, or whether the system will be able to train university mathematics teachers and researchers. At present, practically all PRC mathematicians were trained before the Cultural Revolution, many of them abroad. How will they be replenished? This seems to be a central question.

7.2 MATHEMATICAL EDUCATION IN THE MIDDLE SCHOOLS

Our information on this subject comes from an evening spent with five members of the 47-member team that plans the mathematics curriculum for the primary and middle schools of Shanghai. The team has been together for 9 years, includes three workers and 12 women, and its members range in age from 20 to 71. There are similar teams in other parts of the country, and the various teams communicate with each other by exchanging materials and personnel. Although within broad guidelines there is some local autonomy in developing the mathematics curriculum, the curriculum in most localities is similar to that of Shanghai, we were told. Much of the local variation stems from Chairman Mao's advice to "combine theory with practice," for the "practice" included in the curriculum depends on the needs of local industry or agriculture.

Of the five members of the team with whom we spoke, one had been a middle-school mathematics teacher, two were mathematics graduates of Shanghai Normal College, one was a chemistry graduate of Futan University, and one had been a worker (and member of the propaganda team) in a factory making electrical machinery. Their carefully prepared presentation began, as did all discussions of education that we were to hear in China, with a review of the evils of the "old system," meaning the system used before the Cultural Revolution. (For a detailed discussion of that system and its history, see Frank Swetz, *Mathematics Education in China: Its Growth and Development,* M.I.T. Press, 1974.) In that system, algebra, plane and solid geometry, trigonometry, and analytic geometry were taught as separate subjects, all on a theoretical basis; geometry was taught axiomatically. There was no attempt to explain the practical importance of the material. We heard complaints that "many graduates didn't like to combine with workers and peasants" and that "Chairman Mao's revolutionary line couldn't be implemented." However, after "the Great Proletarian Cultural Revolution destroyed the revisionist line," it was possible to modify education so as to take account of the "three great revolutionary movements": class struggle, the revolution in production, and the revolution in scientific experiment. One of the goals of Chinese education today is to teach the students to "love socialist society, and to love the workers and peasants."

We were told that the current approach to mathematics is based rather explicitly on certain aspects of dialectical materialism, and in particular on the belief that learning proceeds "from practice to general knowledge and again to practice." Engels was quoted a number of times, in particular as saying that "mathematics is supplementary to dialectical materialism, and expresses a form of dialectical materialism." This was illustrated by the "dialectic relation" between positive and negative numbers, powers and roots, constants and variables, and differentials and integrals, all of which are "pairs of opposites that depend on each other."

We have no basis for judging the extent to which the political and philosophical notions above appear explicitly in the teaching of mathematics or in the thinking of the average mathematics teacher. However, in talking to the curriculum team, we did become convinced that these notions play an important part in their approach to their task.

So much time was spent in presenting the political and philosophical background of the mathematics curriculum that there was not much time available for discussing the details of the curriculum. Moreover, since the curriculum is still regarded as experimental, most texts are rewritten every year; despite our requests, it was impossible for us to see copies of the texts. Nevertheless, the following details emerged from the discussion:

1. Mathematics from the first through the tenth year (that is, through the 5 years of primary school and the 5 years of middle school) is taught as a unified course, with no artificial divisions between subjects and with much attention to practical applications. The study is guided by dialectical materialism, by the desire to "combine form and number," and by the desire to "combine theory with practice."

2. The first 2 years of middle school are devoted to basic facts from algebra and geometry. The Pythagorean theorem is proved by area considerations based on subdividing a square. Parallel lines are introduced intuitively in terms of the equality of corresponding angles when cut by a transversal, and equality of the alternate interior angles is then "proved." Similar plane figures are introduced as a basis of measurement and are then applied to actual measurements in surveying, building of bridges and ships, and so on. In general, the practical utility of various geometric figures and constructions (e.g., finding the center of a circle or a circular arc through three given points) is emphasized. The pre-Cultural Revolution axiomatic approach to plane geometry has been largely abandoned, though some aspects of formal logic (as a means of deducing useful properties from other useful properties) are retained in an unsystematic way.

3. Little was said about algebra. The irrationality of $\sqrt{2}$ used to be demonstrated, but this is no longer done. One project, carried out in conjunction with a factory, was to write a program for computer control of a milling machine, and this involved some use of Boolean algebra

and expression of numbers in the binary system. Expression in other bases is also taught. Complex numbers apparently are not mentioned in the middle school.

4. In plane analytic geometry, the discussion of polar coordinates includes the equations of Archimedean spirals and their use in cam designs. The notion of parametric equations of a curve is introduced by means of the cycloid and its evolute, and their uses in various production problems are explained.

5. The third and fourth years of middle-school mathematics are devoted to logarithms, the function concept, and basic concepts of calculus. Natural logarithms are mentioned, but their importance is not explained to the student. Differentiation and integration of polynomials are based on geometric intuition.

6. In the fifth (last) year of middle school, the students learn more calculus, some statistics, and some elementary methods of operations research. These include the use of orthogonal experimental designs, and the use of Fibonacci search or golden section search for optimization of a unimodal function of one variable.

7. There is considerable attention devoted to computation. During their work in factories, students have access to a computer. Some schools have built their own computers.

8. All mathematics training reflects the belief that "although mathematics is abstract in appearance, it originated from the real world and is a product of the needs of the people." Thus, in addition to frequently hearing about practical problems in the classroom, each student spends a month during the school year on a farm or at a factory. He is taught there about applications of mathematics and is encouraged to look for more applications.

There is a second group in Shanghai working on middle-school mathematics. It is centered at Shanghai Normal College and its materials are used in a few schools; all others use the materials from the writing team. They claim a theoretical distinction between the two, hinging on whether synthesis → contradiction → synthesis or contradiction → synthesis is more appropriate.

The following supplementary comments are based on a visit to the May 7th Middle School in Shanghai. (It is at present a 4-year school, though next year it will be a 5-year school.) Its students have a total of 26 45-minute classes per week, and for the fourth-year students, they are as follows: five mathematics, two basic industrial knowledge (including some physics and chemistry), two agriculture (including some biology), five Chinese, four English, two geography, one history, two revolutionary arts and literature (including drawing), and two physical training. There is a 2-week winter vacation and a 6-week summer vacation. All fourth-year students spend 1 month during the school year in agricultural work; all third-year students spend a month in industry and a month in military studies. Examinations are given in the factories and also in classes (twice each term), but they are different from the examinations given "before the Cultural Revolution, when the student took the teacher as his enemy." Now the examinations are often open-book or oral, and they consist of practical problems.

Each teacher gives 10 classes per week. All students follow the same curriculum, which involves four or five mathematics classes per week through all 4 years. We asked about "dropouts" and were told, "It has never happened. All students complete the course."

8

CONCLUSIONS

Our delegation was received most cordially by the individuals and institutions we visited in China. The Institute of Mathematics and other institutes of the Academia Sinica, universities and industries, and our hosts at official functions all provided us with a great deal of information. Our conclusions are based chiefly on these first-hand contacts and therefore may not apply to other mathematical efforts in the People's Republic of China. The mathematical work we saw, both the pure and the applied, is ably and energetically pursued. Its character is described in the previous chapters and, therefore, repetition here would be superfluous.

The size of the mathematical community is surprisingly small for so vast a nation, but with a vigorous educational program, it could grow along with other technological and scientific activities. Chinese mathematicians constitute three distinct generations. There is a small group of mathematicians trained in the West (usually before 1950), a larger generation trained in China and Russia (1950-1966), and the new group of students just now graduating with the first university classes in mathematics since the Cultural Revolution. A few members of the first two generations are still active in pure mathematics, and some of them have achieved results worthy of international recognition. In recent years, members of all three generations have been impressively effective in applying mathematical methods (e.g., operations research and the finite element method) to agriculture and industry.

The lack of systematic training in mathematics concerns us, however, because we know that in many countries the health and vigor of the subject depend on a steady output of new ideas from young Ph.D's. Furthermore, China's current research and training programs in the more applied aspects of mathematics are unlikely to provide an adequate reservoir of understanding and skill for the future. As the country continues to develop, further technological problems--some peculiar to China's situation--are sure to arise. In order to meet the needs of both pure and applied mathematics, it is important that China maintain a strong and varied competence in mathematics as a *discipline*.

The Chinese teachers whom we met were devoted and enthusiastic in searching for new and effective teaching methods. The educational system is still in a state of flux; the changes brought about by the Cultural Revolution may not yet have been assimilated. The number of universities is not large, and there is a dependence on short-term and correspondence courses. At present, many young people do not have much opportunity for advanced study. Approximately 1 percent of middle-school graduates go on to a university. Several universities are just beginning to institute graduate work.

Both in research and in education, current Chinese doctrine emphasizes the importance of combining theory with practice, and of open-door research directed to problems arising from the experiences of workers and peasants. These dicta have predominated over other motivations, such as simple curiosity or joy derived from the beauty and symmetry of mathematical ideas. Generally, in the past, all of these motivations have combined to encourage mathematical development. Mathematics emerging from the purest of beginnings has frequently spawned very powerful tools for science and technology. Even from a strictly utilitarian viewpoint, it seems certain that a policy of additional encouragement of broader mathematical research and education would serve China well.

In any country, there is an inherent danger that research policies will be set by professional managers who have not themselves done first-rate creative work. The present Chinese emphasis on the "open-door" and "3 in 1" approaches to research, education, and applications sometimes places technically under-qualified persons in leadership and decision-making roles. Such a situation could lead in the future to low standards and could discourage real creativity. Although the work we saw was very good on the whole, we feel that this phenomenon could be especially troublesome in a country like China, where national priorities require an emphasis on repeated applications of known techniques. The best protection is the development of an independent group of first-rate scientists and mathematicians.

The elimination of elites, a primary goal of the Cultural Revolution, has fostered a widely stated corollary that talent is shared about equally by all. This view is at variance with competing notions of the importance of exceptional talent and of the need to recruit especially gifted young people to maintain vigor in mathematics. Our own experience suggests that some combination of talent, motivation, and interest selects a few people who are especially effective in mathematics (as also in sports, art, or politics).

Our trip to China has been a very positive experience. All of us learned a great deal about China, and, for many of us, what we heard and saw has prompted an analysis of our own assumptions. This has "raised our consciousness" on a number of issues, such as the following:

1. In our concern for the mathematically talented student, are we neglecting those students who, while not especially gifted, will be users of mathematics?
2. Should we, in our teaching and writing, pay more attention to the relation of mathematics to other sciences, engineering, philosophy, history, and economics?
3. To what extent should academic mathematicians combine theory and practice by engaging in industrial, government, and military research?
4. To what extent should motivation through applications be available throughout the mathematics curriculum?

The members of the Mathematics Delegation hope that our visit will contribute to the establishment of close relations between Chinese and American mathematicians. We look forward to the privilege of hosting a Chinese mathematics delegation soon. We are confident that these visits will usher in an active exchange program both in research and in teaching. To this end we urge the American mathematical community and our government officials to make every effort to increase contacts with our Chinese colleagues. We hope that these contacts will assist both the United States and China.

APPENDIX A

BRIEF ITINERARY OF THE PURE AND APPLIED MATHEMATICS DELEGATION:

MAY 3-27, 1976

May 3-17	PEKING
May 4	Institute of Mathematics
May 5	Institute of Mathematics Welcoming banquet hosted by Chou P'ei-yüan, Vice Chairman of the Scientific and Technical Association (eve)
May 6	Institute of Mathematics (Group 1) Diesel Locomotive Depot and Institute of Mathematics (Group 2) Lecture by Wu Chung-hua, Vice-Director of Institute of Mechanics and informal discussions (eve)
May 7	Peking University Lectures by four members of Institute of Mathematics (eve)
May 8	Peking University Luncheon hosted by Hua Lo-keng at the Summer Palace Lectures by U.S. delegation members at Peking University Informal discussion with mathematicians from Institute of Mathematics and Peking University in hotel (eve)
May 9	Lectures by scientists from Institute of Oceanography (Carrier) People's Printing Machinery Plant (Group 2) Imperial Palace and Palace Museum (p.m.) Discussion in hotel with members of Institute of Oceanography (eve) Lecture in hotel by Keller (eve)
May 10	Institute of Meteorology (Group 1) Feng T'ai Car Insulation Depot (Group 2)

	Lectures by Kohn and Klee at Institute of Mathematics (Group 3) Tsinghua University (p.m.) Discussion in hotel with members of Institute of Biophysics (Keller)
May 11	Institute of Mathematics
May 12	Institute of Mathematics Visit to hotel by mathematicians from Institute of Mathematics (eve)
May 13	Institute of Mathematics--lectures by Mac Lane, Brown, and Feit
May 14	Institute of Mathematics

<u>May 11-15 Subgroup to HARBIN</u>

	Visits to: Tach'ing Oil Field, Tongkiang Transistor Factory, and Steam Turbine Plant Lectures on applied mathematics in hotel Discussions with faculty from Heilungkiang University and Harbin Industrial College Banquet hosted by Ts'ao Chih, Vice-Chairman of the Revolutionary Committee of Heilungkiang Province
May 15	Great Wall and Ming Tombs (Group 1) Reception at U.S. Liaison Office (Both groups, eve)
May 16	Great Wall and Ming Tombs (Group 2) Thank-you banquet hosted by U.S. delegation (eve)
May 17	Institute of Mathematics--lectures by Keller and Pollak Historical Museum (Group 2) Delegation received in the Great Hall of the People by Yao Lien-wei, Vice-Chairman of the Standing Committee of the National People's Congress

<u>May 18-19 Travel to and sightseeing in SOOCHOW</u>

Banquet hosted by Ms. Wang, Vice-Chairman of the Soochow Municipal Revolutionary Committee (eve)

<u>May 20-27 SHANGHAI</u>

May 20	Futan University

May 21	Futan University Banquet hosted by Yang Hsi-kuang, Vice-Chairman of the Shanghai Municipal Revolutionary Committee (eve)
May 22	Futan University--including lectures by U.S. delegation members Volleyball game between Swedish and Chinese teams (eve)
May 23	May 7th Cadre School for Scientific Workers "White-haired Girl" (eve)
May 24	Shanghai Hua-tung Normal University Thank-you banquet hosted by U.S. delegation (eve)
May 25	Shanghai Hua-tung Normal University--lectures by U.S. delegation (Group 1) Shih P'ing Workers' Village (Group 2) Shanghai Hua-tung Normal University (Both groups, p.m.) Visit to hotel by Shanghai Teaching Materials Compiling Group (eve)
May 26	Talks by Mac Lane, Carrier, and Pollak at Science and Technology Center (Group 1) May 7th Middle School (Group 2) Malu Commune (p.m.) Shanghai acrobatics performance (eve)
May 27	Shanghai Industrial Exhibition Sightseeing in Old City Depart for Tokyo

APPENDIX B

CHINESE NAME LIST

Delegation's Escorts

苏风林 Su Feng-lin
Staff Member
Bureau of Foreign Affairs
Scientific and Technical Association

桂湘雲 Kuei Hsiang-yün
Institute of Mathematics
Operations Research

戴新生 Tai Hsin-sheng
Institute of Mathematics
Topology

朱培华 Chu P'ei-hua
Institute of Mathematics
Member of Planning Bureau and Interpreter

朱世学 Chu Shih-hsüeh
Institute of Mathematics
Interpreter

Hosts in Peking

周培源 Chou P'ei-yüan
Vice-Chairman
Scientific and Technical Association

刘华清 Liu Hua-ch'ing
Leading Member
Chinese Academy of Sciences

朱永行 Chu Yung-hang
Deputy Director
Foreign Affairs Bureau
Scientific and Technical Association

馮因复

Feng Yin-fu
Deputy Chief of Division
Foreign Affairs Bureau
Scientific and Technical Association

錢 皓

Ch'ien Hao
Staff Member
Bureau of Foreign Affairs
Scientific and Technical Association

任南衡

Jen Nan-heng
Staff Member
Scientific and Technical Association

胡风仙

Hu Feng-hsien
Department of American and Oceanian Affairs
Ministry of Foreign Affairs

Special Host

姚連蔚

Yao Lien-wei
Vice-Chairman of Standing Committee
National People's Congress

INSTITUTE OF MATHEMATICS

General

华罗庚

Hua Lo-keng
Director

趙蔚山

Chao Wei-shan
Chairman
Revolutionary Committee

田方增

T'ien Fang-tseng
Vice-Director

賈雨文

Chia Yü-wen
Member
Revolutionary Committee

罗声雄

Lo Sheng-hsiung
Member
Revolutionary Committee and Planning Bureau

Number Theory

陳景潤 Ch'en Ching-jun

王 元 Wang Yüan

Algebra

刘木兰 Liu Mu-lan

万哲先 Wan Che-hsien

Topology

張素誠 Chang Su-ch'eng

戴新生 Tai Hsin-sheng

王启明 Wang Ch'i-ming

吴文俊 Wu Wen-tsün

Analysis

張广厚 Chang Kuang-hou

馮 康 Feng K'ang

韓天敏 Han T'ien-min

許以超 Hsü Yi-ch'ao

杨 乐 Yang Lo

Ordinary Differential Equations

陳兰孙 Ch'en Lan-sun

秦元勳 Ch'in Yüan-hsün

Partial Differential Equations

张　同	Chang T'ung
顧永耕	Ku Yung-keng
李孝貴	Li Hsiao-kuei
丁夏畦	Ting Hsia-hsi
王光寅	Wang Kuang-yin

Operations Research

朱永津	Chu Yung-chin
韓継业	Han Chi-yeh
徐瑞恩	Hsü Jui-en
徐光輝	Hsü Kuang-hui
徐国志	Hsü Kuo-chih
顧基发	Ku Chi-fa
桂湘雲	Kuei Hsiang-yün
刘彦佩	Liu Yen-p'ei
馬仲蕃	Ma Chung-fan
董直清	Tung Chih-ch'ing
越民义	Yüeh Min-yi

Probability and Statistics

安鴻志	An Hung-chih
章照止	Chang Chao-chih
陈培德	Ch'en P'ei-te
成 平	Ch'eng P'ing
項可风	Hsiang K'o-feng

Control Theory

陈汉馥	Ch'en Han-fu
秦化淑	Ch'in Hua-shu

Computer Science

陸汝鈐	Lu Ju-ch'ien
邵秀民	Shao Hsiu-min

Other

江嘉禾	Chiang Chia-ho
肖 玲	Hsiao Ling
李泙榮	Li P'ing-jung
陸柱家	Lu Chu-chia
枚明澂	Mei Ming-ch'eng
沈信耀	Shen Hsin-yao
王請华	Wang Ch'ing-hua

吴新谋 Wu Hsin-mou

余燕林 Yü Yen-lin

INSTITUTE OF MECHANICS

吴仲华 Wu Chung-hua, Deputy Director

张保楝 Chang Pao-tung

吴文权 Wu Wen-ch'üan

余申 Yü Shen

INSTITUTE OF OCEANOGRAPHY

方国洪 Fang Kuo-hung

任允武 Jen Yün-wu

顧玉荷 Ku Yü-ho

毛汉礼 Mao Han-li

INSTITUTE OF BIOPHYSICS

张少吾 Chang Shao-wu
Mathematics of compound eyes

郑竺英 Cheng Chu-ying
Binocular vision

刁云程 T'iao Yün-ch'eng
Physiology of vision

汪云九 Wang Yün-chiu
Mathematical models of receptor field

PEKING UNIVERSITY

黄辛白 Huang Hsin-pai
Vice-Chairman
Revolutionary Committee

周俊业	Chou Chün-yeh Responsible Member Revolutionary Committee
李秀茹	Li Hsiu-ju Member, Office of Revolutionary Committee
陈守良	Ch'en Shou-liang Chief, Natural Science Office
段学复	Tuan Hsüeh-fu Professor and Chairman Revolutionary Committee of Mathematics Department
张芷芬	Chang Chih-fen Associate Professor and Vice-Chairman Revolutionary Committee of Mathematics Department
廖山涛	Liao Shan-t'ao Professor of Mathematics
张恭庆	Chang Kung-ch'ing Teacher, Mathematics Department
黄永念	Huang Yung-nien Teacher, Mathematics Department
郭仲衡	Kuo Chung-heng Teacher, Mathematics Department
廖可人	Liao K'o-jen Teacher, Mathematics Department
应隆安	Ying Lung-an Teacher, Mathematics Department
陈新	Ch'en Hsin Teacher of English
王武仁	Wang Shih-jen Teacher of English
何晓田	Ho Hsiao-t'ien Student Mathematics Department
李学增	Li Hsüeh-tseng Student Mathematics Department

李紹孟 Li Shao-meng
Student
Mathematics Department

王淑芬 Wang Shu-fen
Student
Mathematics Department

王东迁 Wang Tung-ch'ien
Student
Mathematics Department

TSINGHUA UNIVERSITY

张維 Chang Wei
Vice-Chairman
Revolutionary Committee and
Professor of Civil Engineering

馬文仲 Ma Wen-chung
Vice Head
Administrative Office of Revolutionary
Committee and Teacher of Physics

錢偉長 Ch'ien Wei-ch'ang
Professor of Mechanics

趙訪熊 Chao Fang-hsiung
Professor of Mathematics

张万琦 Chang Wan-ch'i
Teacher of Mathematics

柳西玲 Liu Hsi-ling
Teacher of Mathematics

PEKING DIESEL LOCOMOTIVE DEPOT

李国富 Li Kuo-fu
Vice-Chairman
Revolutionary Committee
Peking Internal Combustion
Engine Service Department

霍洪奎 Huo Hung-k'uei
Chairman
Disassembly Workshop
Peking Internal Combustion
Engine Service Department

张崇华
Chang Ch'ung-hua
Worker, Peking Internal Combustion Engine Service Department

李学华
Li Hsüeh-hua
Foreign Affairs Worker
Peking Internal Combustion Engine Service Department

李国樑
Li Kuo-liang
Foreign Affairs Worker
Peking Railroad Administration

吴 潤
Wu Jun
Foreign Affairs Worker
Peking Railroad Administration

Hosts in Harbin

曹 志
Ts'ao Chih
Vice-Chairman
Revolutionary Committee
Heilungkiang Province

張維信
Chang Wei-hsin
Vice-Chairman
Committee on Science and Technology
Heilungkiang Province

李明文
Li Ming-wen
Responsible Member
Committee on Science and Technology
Heilungkiang Province

龙海涛
Lung Hai-t'ao
Division Chief
Committee on Science and Technology
Heilungkiang Province

張树德
Chang Shu-te
Staff member
Committee on Science and Technology
Heilungkiang Province

吴文东
Wu Wen-tung
Vice-Chairman
Hua-nan Hsien Ta-pa-ling Commune

Mathematicians in Harbin

宋保生 Sung Pao-sheng
Worker-teacher
Ta-hai Forestry Bureau
Heilungkiang Province

杜长泰 Tu Ch'ang-t'ai
Vice-Chairman
Mathematics Department
Heilungkiang University

刘家琦 Liu Chia-ch'i
Teacher of Computational Mathematics
Harbin Industrial University

刘大昕 Liu Ta-hsin
Teacher of Computational Mathematics
Harbin Industrial University

奚盤根 Hsi P'an-ken
Shantung Oceanographic College

葛有信 Ko Yu-hsin
Shantung Oceanographic College

馮士筰 Feng Shih-tso
Shantung Oceanographic College

王賀臣 Wang Ho-ch'en
Vice-Chairman
Revolutionary Committee
Harbin Steam Turbine Plant

Shanghai Hosts

杨西光 Yang Hsi-kuang
Vice-Chairman
Municipal Revolutionary Committee of Shanghai

許言 Hsü Yen
Responsible Member, Shanghai Scientific and Technical Association

毛生芽 Mao Sheng-fang
Responsible Member, Administrative Office, Shanghai Scientific and Technical Association

朱汉达 Chu Han-ta
Secretary, Shanghai Scientific and Technical Association

胡敏民 Hu Min-min
Secretary, Shanghai Scientific and Technical Association

陸关虎 Lu Kuan-hu
Secretary, Shanghai Scientific and Technical Association

FUTAN UNIVERSITY

Keng Lien-nien
Vice-Chairman
Revolutionary Committee

陈秩寿 Ch'en Chih-shou
Educational Revolution Group

曹运揮 Ts'ao Yün-hui
Educational Revolution Group

江风記 Chiang Feng-chi
Responsible Person
Mathematics Department

苏步青 Su Pu-ch'ing
Professor of Mathematics

谷超豪 Gu Ch'ao-hao
Differential Geometry

夏道行 Hsia Tao-hsing
Functional Analysis

舒文昌 Shu Wen-ch'ang
Functional Analysis

严紹宗 Yen Shao-tsung
Functional Analysis

許永华 Hsü Yung-hua
Algebra

陈恕行 Ch'en Shu-hsing
Differential Equations

金福臨 Chin Fu-lin
Differential Equations

胡和生 Hu Ho-sheng
  Differential Equations

侯志伸 Hou Chih-shen
  Mathematics Teacher

忻元新 Hsin Yüan-hsin
  Mathematics Teacher

华宣积 Hua Hsüan-chi
  Mathematics Teacher

李訓經 Li Hsün-ching
  Mathematics Teacher

李大灒 Li Ta-tsan
  Mathematics Teacher

柳兆榮 Liu Chao-jung
  Mathematics Teacher

吕慧芳 Lü Hui-fang
  Mathematics Teacher

欧阳鬯 Ou-yang Yü
  Mathematics Teacher

沈純理 Shen Ch'un-li
  Mathematics Teacher

施嗣伯 Shih Ssu-po
  Mathematics Teacher

吴立德 Wu Li-te
  Mathematics Teacher

俞文魮 Yü Wen-chi
  Mathematics Teacher

馮善萍 Feng Shan-p'ing
  English Teacher
  Interpreter

竺曼莉 Chu Man-li
  Student

何品鈞 Ho P'in-chün
  Student

李志远 Li Chih-yüan
  Student

王 健 Wang Chien
Student

吴偉祖 Wu Wei-ch'ang
Student

袁银华 Yüan Yin-hua
Student

SHANGHAI HUA-TUNG NORMAL UNIVERSITY

万东輝 Wan Tung-hui
Responsible Member
Revolutionary Committee

林火土 Lin Huo-t'u
Vice-Chairman
Revolutionary Committee
Mathematics Department

曹錫华 Ts'ao Hsi-hua
Professor of Applied Mathematics

陈志杰 Ch'en Chih-chieh
(N. A.)

蒋金林 Chiang Chin-lin
Worker-lecturer
Mathematics Department

蒋師付 Chiang Shih-fu
Worker-lecturer
Mathematics Department

杭永珍 Hang Yung-chen
Computer Mathematics

胡啓迪 Hu Ch'i-ti
Mathematics Teacher

李銳夫 Li Jui-fu
Professor, History of Mathematics

孙莱祥 Sun Lai-hsiang
Mathematics Teacher

袁德东 Yüan Te-tung
(N. A.)

張永祺 Chang Yung-ch'i
(Lectured on rice transplanter)

蒋偉成 Chiang Wei-ch'eng
(Lectured on cables)

吴文权 Wu Wen-chüan
(Lectured on finite element method)

刘干成 Liu Kan-ch'eng
Mathematics Device Manufacturing Team

黄榮荃 Huang Jung-ch'üan
Research in Middle-School Education

郑啓明 Cheng Ch'i-ming
Revolutionary Education Team

胡应平 Hu Ying-p'ing
Revolutionary Education Team,
Mathematics Department

茆詩松 Mao Shih-sung
Revolutionary Education Team,
Mathematics Department

童友德 T'ung Yu-te
Teacher, Workers' Propaganda Team

朱水林 Chu Shui-lin
Natural Dialectic Theory

吴炳榮 Wu Ping-jung
Natural Dialectic Theory

陈德金 Ch'en Te-chin
Responsible Member, Revolutionary Committee of School-Managed Factory

李巧云 Li Ch'iao-yün
Student

潘立明 P'an Li-ming
Student

SHANGHAI TEACHING MATERIALS COMPILATION GROUP

孙元清 Sun Yüan-ch'ing
Responsible Person

陈肇曾 Ch'en Chao-tseng

周以剛 Chou Li-kang

馮 賢 Feng Hsien

华德清 Hua Te-ch'ing

APPENDIX C

NON-ACADEMIC VISITS

C.1 VISIT TO TACH'ING OIL FIELD

Throughout China one constantly sees and hears the slogan, "Industry [should] learn from Tach'ing." Our visit to the great Tach'ing oil field, though only for a day, gave our delegation the opportunity to sense the spirit behind that ubiquitous slogan.

The Tach'ing oil field is situated in Heilungkiang Province, China's most northeasterly province, only about 100 km from the USSR border. Before 1959 the area was a vast, underdeveloped, and sparsely populated grassland. On September 26, 1959, the first well of the field, No. 3, was brought in. Because that date was the eve of the tenth anniversary of the founding of the People's Republic of China, the new field was named Tach'ing, meaning "Great Congratulations." This initial success came only 6 months after promulgation of an official government policy to develop China's natural resources. By 1963, the Tach'ing field had made China self-sufficient in oil, and in 1964 Chairman Mao issued his famous call: "Industry [should] learn from Tach'ing."

The significance of the slogan is made manifest in many ways at Tach'ing, now a kind of shrine from which China's industrial pilgrims draw inspiration. The quintessential feature exemplified by Tach'ing is "self-reliance," another major slogan with a long historical tradition. The aim of self-reliance is to accomplish major industrial development--with minimal financial or technical aid--by application of the "genius of the masses," a virtually limitless use of physical labor, and Chairman Mao's ideological guidance.

The "Iron Man Wang Chin-hsi Memorial Hall" honors an exemplary individual, the self-reliant spirit behind the Tach'ing slogan. It is a museum filled with the tools, clothing, books, and other memorabilia of the model worker-cadre Wang Chin-hsi. The stories and photographs of him imply that Wang led the way, single-handedly, in creating this oil field, by dint of his will, hard work, and application of Mao Tse-tung Thought. A marvelously detailed animated model of well No. 3 shows how the workers, lacking pumps or pipes to supply water for their drill, set up a lengthy "basin brigade" to pass water by hand from a distant pond. Bright-eyed girl guides relate these stories in worshipful and breathless tones while the miniature drill bores heroically into the miniature steppe.

The reality of Tach'ing is sufficiently impressive not to require propagandistic embellishment. The production of the field is a secret, but a model well we visited had been producting 70 bbl./day for 16 years; with the landscape dotted for miles in every direction with white-painted well-huts on a grid of 250 meters, one surmises that production must be prodigious indeed. We were told, furthermore, that production has regularly increased 30 percent/annum by such methods as forcing water into the wells to maintain pressure so that no pumping is required. Crude output today is six times what it was in 1965, and the drilling of new wells continues. Drilling teams compete for total depth drilled annually; the record of 127,000 meters was set in 1971 by team No. 1205.

Besides crude petroleum production, the Tach'ing fields include processing stations (especially to dry the oil), refineries, and petrochemical industry. A refinery we visited, the Tach'ing General Petrochemical Factory, was established in 1962 and began production in 1963 with about 1 million tons annually of 18 different products. It now produces about 5 million tons annually of 45 different products.

In keeping with the spirit of self-reliance, even the small investment of state capital has been repaid 13 times over. The total value of the refinery/petrochemical plant mentioned above is now 11 times the original investment. The families of oil workers have engaged in agricultural production, and over the last 16 years they have produced 500 million kg of grain and 700 million kg of vegetables; they have taken no state grain at all for the past 11 years.

In addition to its 60 settlements housing peasants and 100 housing workers, the Tach'ing oil field maintains shopping centers, one "oil college," 70 "workers' universities," 250 schools at all levels, and 150 health centers and hospitals.

C.2 THE MAY 7TH CADRE SCHOOL FOR SCIENTIFIC WORKERS--SHANGHAI AREA

Since its founding in January 1969, this May 7th Cadre School has trained 8,000 persons: 2,100 party cadres and 5,900 scientists. Of these, 27 percent were women. There are usually about 500 students in residence and a staff of about 160, of whom 50 to 60 are full-time workers. The training program lasts 3 months, with 3 days of vacation each month. Four hours a day are allotted to the study of Marxism-Leninism and Mao Tse-tung Thought, and 4 hours to physical labor. The school grows vegetables and grain, and manufactures bricks. It appears to be prosperous.

The students' daily schedule is as follows:

5:30 a.m.	reveille
5:50	exercise, housekeeping
6:15	breakfast
7:00-11:30	study, labor
11:30	lunch
12:00-2:00 p.m.	rest
2:00-5:30	study, labor
5:30	supper

7:00-8:30	recreation, free
9:00	lights out

The regimen of the school does not appear to be too onerous, and the students seem to enjoy it. The political discussions do not include formal self-criticism sessions, and much of the heavy work seems to be done by the professional workers.

Attending the school is considered very desirable and many people apply. There is usually a long waiting period. People who wish to come a second time must wait at least 7 years before returning. The age limit is 50 and the people admitted must be in good health. There are about 40 such schools that serve Shanghai and about 90 percent of the personnel in scientific institutes attend them.

Our delegation spent a fine spring Sunday traveling from Shanghai through the flat farmlands in order to visit this school. We were well received, as always. We walked through the fields to see the students at their labors, and we then were treated to poetry and song and were served a bounteous meal. Afterwards we talked with a selected group of men and women attending the school. They were evidently enjoying their stay. They had left children and family back in Shanghai to take part in a new experience, and they enjoyed the change.

We were left with the impression that attending this school is a very positive experience. Most of the activities there are pleasant; there is a good "esprit de corps" and attendance is considered meritorious in bringing the intellectuals closer to the working class.

APPENDIX D

EXAMPLES OF PRESENTATIONS TO THE DELEGATION

D.1 SOME ACCOMPLISHMENTS OF OUR INSTITUTE IN APPLIED MATHEMATICS: A TRANSLATION OF REMARKS BY CH'ENG P'ING AT THE INSTITUTE OF MATHEMATICS

The research personnel of our Institute, having gone through the Great Proletarian Cultural Revolution, the movements to criticize Lin Piao and Confucius, and most recently, to criticize Teng Hsiao-p'ing and counter-attack the right deviationist wind reversing correct verdicts of the Cultural Revolution, have reaffirmed their will to implement completely the guidelines of the Constitution "to serve proletarian politics, to serve the workers, peasants and soldiers, and to join in productive labor," and to do mathematical research according to the principle of "Practice-Theory-Practice." To this end we employ the method of "open-door" scientific research.

We have changed the past method of mathematical research: sitting in an office working from document to document, with a pen and a cup of tea. We term that the "three separations," that is, separation from politics, separation from workers and peasants, and separation from reality (practice). Now we go out among the workers and peasants to receive re-education, to learn from them, absorb nourishment, select research topics according to their needs, and carry out research according to their experience. In return, we give them some mathematical methods to mobilize the broad masses of workers, peasants and soldiers to do research together under the leadership of the Party. We call this "Three-in-one," that is workers, cadres and technicians or research personnel, all three together. Factories, schools, and scientific research units also carry on their work according to the "Three-in-one" principle. For the past few years we have been engaged, on the whole, in the following kinds of work:

I. Our mathematical workers, in combination with workers and peasants according to their demands, have simplified some mathematical methods and have used the commonly understood language of the masses to propagate the concepts of dialectical materialism. Accordingly, the broad masses of workers, peasants, and soldiers have employed the "optimal seeking method," the method of orthogonal design, scheduling methods, and so on, in extensive scientific experiments to resolve problems in various trades and industries. On a national basis, an enormous number of results have been achieved, and our audience has been very extensive. In trade and industry, we have had comparatively good results: in agriculture, with such items as comparison of seed breeding and fertilizer application, prescribing fodder and agricultural chemicals, allocating labor, etc.; in industry, in such areas as chemical industry, light industry, weaving, machinery, metallurgy, electronics, pharmaceuticals, sanitary food products, railroads, construction, etc. We have also been able to achieve results in increasing the speed of testing and engineering, economizing raw materials, breaking technical bottlenecks, and aiding the search for rules of engineering techniques. In the process of extensively utilizing these applications, there arose a great many items still awaiting study, and these supplied a source of further mathematical development.

For example, the problems of group testing of the optimal seeking method, of optimization with time delay, and of unconstrained and constrained optimization methods promoted further research on optimal design. Our comrades have already obtained some results. In excellent research on the method of orthogonal design, it was felt that past theory was now insufficient to explain results in today's practice. We are presently opening a study of this question. The noteworthy experience in scientific experiment of the worker-peasant-soldier masses supplies important lines and principles for our thinking on these questions. The special points of our expanded work are described below:

1. Workers-peasants-soldiers manifest the function of the main force in scientific experiment. Their normal experimentation, however, suffered from lack of scientific method to guide them and a resultant period of groping about. Now that they have grasped these tools and combined them with their own experience, they are able to determine proposals together with technical personnel, and their common experimental power is then greater. One example is in scheduling methods. Since workers are the ones who determine plans and implement them, good results may be obtained if all personnel participating in the work clearly understand the work procedures and their own responsibilities.

2. The majority of provinces and municipalities have established special organs to manage projects planned under Party leadership.

3. The use of Marxist-Leninist philosophy for guidance and explanation enables the masses to understand easily. For example, the optimal seeking method stresses grasping "the main contradiction," and the method of orthogonal design stresses that substances are formed of many facets and complex factors, so that people must draw out the most important points from the broad relationships. Thus they study both philosophy and mathematics.

II. We elevate the experience and creative invention of the masses to be our new theory. When we Chinese first began to study operations research, our mathematical workers went out and studied the methods of shipping grain used by grain offices, and proved theoretically that it was already optimal. We called this "algorithms in graph theory" and carried it further to raise what we term in ordinary documents "The Chinese Postman Problem." In the last 2 years, moreover, we discovered that the railroad offices had a good method of allocating cars, and that provided theoretical research topics for which we have already worked out partial results. Together with our institute's homemade stand-alone computer and programs compiled to expand its use in practice, this will, I believe, enable us to promote the use of computers to control traffic on China's railroads.

III. The great problems raised in socialist construction require us mathematical workers to render assistance and to join in solving them.

1. For problems raised in geological exploration and mining, the question is how numerically to resolve problems of storage capacity, placement, classification of routes, and processes of loading and transport in mines. Mathematical statistics, modeling and differential equations are important tools for solving these types of problems. Comrades in our Institute started from the Cauchy problem of the Laplace equation to study problems of geologic exploration, which finally became a problem of nonlinear design. Others used multivariate analysis to obtain very good results.

2. For certain work in determining national standards, we used statistical methods, such as sampling techniques and statistical analysis tools, to aid the offices involved in determining the standards, such as sampling standards for basic electronic parts, assembly standards, standards for analyzing components of steel, and reliability standards. In these we accelerated the study of sampling methods, multivariate analysis, and regression analysis.

3. In problems raised by industrial controls, what especially aroused our interest in control theory was the use of computers in providing automatic controls for industry. Our main tool was linear wave filters, such as Kalman wave filters--a question which aroused our interest. Besides this we proceeded to study damping of elastic vibration, stochastic observability, and satellite attitude control. Some results have been published in the journals *Scientia Sinica (Chung-kuo k'o-hsüeh)* and *Acta Mathematica Sinica (Shu-hsüeh hsüeh-pao)*.

4. Various types of statistical forecasting, such as those for weather, earthquakes, and hydrography, have for these last few years engaged the interest of China's mathematical workers. In our Institute, research in this area has been rather slight, with only individual comrades involved, mainly using the method of steady state theory to make predictions.

5. We have also been working on mathematical problems raised in physics and engineering technology, such as qualitative problems in these ordinary differential equations, which occur in phase-locked loops.

This then is what I have to add to the material reported yesterday by Mr. Chao and Mr. T'ien.

D.2 INTRODUCTION TO MATHEMATICAL TALK BY SU PU-CH'ING: FUTAN UNIVERSITY, SHANGHAI

Students from before the Cultural Revolution can say much in theory but cannot solve any practical problems. There is much mathematical knowledge in practical projects; teachers can learn much from the workers, peasants, and soldiers. Theory linked with practice is the only correct way. I started research in mathematics in the 1920's in differential geometry. I wrote more than 100 papers having nothing to do with practice. I did not know what was useful. What I knew was of no help in practice. It is all out of date now. What you need to do is to begin with a practical problem, dig deeper, and bring theory back to practice. Our country is not advanced enough; if I do *this* kind of work, it is of some help to our country.

Let me reflect about combinations of the old, the middle-aged, and the young. Young people are full of vitality; our hopes for the future, including the future of the mathematics department, are dependent on them. The facts are like this. An old man like me is too old to work. Middle-aged teachers have more new ideas and must help the young teachers along. Old teachers like me cannot say we are of no use at all; we still have some ability. I can play a small part in scientific research by collaborating with young and middle-aged teachers. In the past, I was divorced from practice, from the workers, peasants, and soldiers, and also from young and middle-aged teachers. In the last 3 or 4 years I have participated in open-door schooling, and have done some mathematics by relying on the workers and on young and middle-aged teachers. In one sentence--even if I am in very old age, I must learn actively and act to remold my work continuously.

D.3 DIALOGUE ON THE BEAUTY OF MATHEMATICS: FROM A DISCUSSION AT THE SHANGHAI HUA-TUNG UNIVERSITY

Kohn: Should you not present beauty of mathematics? Couldn't it inspire students? Is there room for the beauty of science?

Answer: The first demand is production.

Kohn: That is no answer.

Answer: Geometry was developed for practice. The evolution of geometry could not satisfy science and technology; in the seventeenth century, Descartes discovered analytical geometry. He analyzed pistons and lathes and also the principles of analytical geometry. Newton's work came out of the development of industry. Newton said, "The basis of any theory is social practice." There is no theory of beauty that people agree on. Some people think one thing is beautiful, some another. Socialist construction is a beautiful thing and stimulates people here. Before the Cultural Revolution some of us believed in the beauty of mathematics but failed to solve practical problems; now we deal with water and gas pipes, cables, and rolling mills. We do it for the country and the workers appreciate it. It is a beautiful feeling.

APPENDIX E

SAMPLES OF THE CONTENTS OF MATHEMATICAL JOURNALS AND BOOKS

E.1 *ACTA MATHEMATICA SINICA* (VOLUMES PRESENTED TO THE DELEGATION BY THE MATHEMATICAL INSTITUTE, PEKING)

Vol. 17, No. 1 (1974)

	pages
Thoroughly criticizing the principles of Confucius Major criticism subgroup, differential equations group of the Institute of Mathematics	1-4
Statistical predictions of earthquake movements Wang Ch'i-ming *et al.* (5 others)	5-19
On certain time series and their energy transport properties Hsu Li-hua	20-27
Homology classes associated with singularities of the S_k type Wu Wen-tsün	28-37
Investigation in the theory of approximation in optimal filtering theory An Hung-chi	38-45
The structure of Q-matrices and the reducibility properties of polynomials over finite fields Tu Kuei-chang	46-59
Certain scattering problems connected with indefinite metrics Hsia Tao-hsing	60-75

Vol. 17, No. 2

From practice comes real knowledge--a critique of the "theory of genius" in idealistic philosophy Chao Hsin-hua & Li Huo-chen	77-78
On the determination of the correct amount of water to be pumped into oil fields by operations research--preliminary report Tach'ing Oil Fields Exploration Institute *et al.*	79-88
On the computation of the natural frequency of a beam Li Ta-chin & Li Li-kang	89-99
Canonical differential systems Liao San-tao	100-109

Vol. 18, No. 3

Vol. 18, No. 4

E.2 BULLETIN OF FUTAN UNIVERSITY, 1975

E.3 MATHEMATICS: ITS COGNITION AND PRACTICE

1975, Vol. 1

E.4 EIGHT MATHEMATICS TEXTS

The Controllability and Observability of Linear Systems (K'uan Ch'ou-chi and Ch'en Han-fu, Control Theory Lab, Institute of Mathematics, Peking, 1975)

1. The Controllability and Observability of a Dynamic System
   - 1.1 Introduction
   - 1.2 The controllability of a linear system
     Appendix
   - 1.3 System structure and controllability
   - 1.4 The complete observability of a linear system
   - 1.5 System structure and observability
   - 1.6 The controllability and observability of a discrete time system
   - 1.7 Examples of the controllability of a dynamic system
2. The Controllability and Observability of a Steady System
   - 2.1 The determination of controllability and observability of a steady system
   - 2.2 The determination of controllability and observability by Jordan norm
   - 2.3 The standard structure of a steady system
   - 2.4 Relations between the transitive function and the controllability and observability in a system
   - 2.5 Discrete steady system
   - 2.6 Examples of the controllability and observability of a steady system
3. Index of Controllability and Measures of Controllability
   - 3.1 Index of controllability and index of observability
   - 3.2 Measures of controllability
   - 3.3 Applications for the measures of controllability

Applications of Mathematics in Industry and Agriculture (Education Department of the City of Peking, 1975, for people in the factory and the field)

1. The Dam of Huai Mountain

First, they find out the measurements of the land. Then they determine the location for the dam such that the amount of work (filling lower places and digging higher places) is minimized.

2. Design of Terraced Fields

In order to maximize the area of planting, it is desirable to design terraced fields. The zero-line (the place where no digging or filling is required) is determined so that the amount of soil dug out is equal to the amount of soil that is needed to fill in.

3. Estimates of the Production of Grain

They estimate how much grain is in a field of unit area by the following:

(1) How many plants in one unit:
   (a) The average distance between two rows
   (b) The average number of plants per unit of length
(2) Average number of seeds per plant
(3) Average weight per seed

4. Application of Functions and Graphs

They explain how to illustrate some functions by a graph, e.g., relations between the speed of the flow and the water level of a river.

5. Experimental Formula for Flows

They give a straight line approximation for some experimental data.

6. Problems on Arch Bridges

They indicate that the shape of an arch bridge is similar to a hyperbola.

7. Calculations of Materials for Building a House

The size of a brick is given. They determine the amount of bricks for building a house from its measurements.

8. The Ingredients for Iron

Experimental data.

9. Applications in Whitening

Experimental data.

10. Small Circles Enclosed in a Big Circle

Consider a circle that encloses four small circles of the same size. Find the maximum radius of the small circles.

11. How to Draw a Cam

12. A Problem on Folding Beds

A design is presented for a folding bed such that the space for storage is minimized.

13. The Angle of a Part of a Tool

Elementary geometric problem.

14. Calculation of the Radius of a Circle from Certain Data

Elementary geometric problem.

Advanced Mathematics (Sian University, Mathematics Group in Department of Communications, 1975)

<u>Chapter 1.</u> <u>How to Study Calculus?</u>

Section 1. Variables and functions

- (a) Constants and variables
- (b) Concepts of functions
  - (1) Examples of functions
  - (2) Definition of a function
  - (3) Variations of a function
- (c) Elementary functions
- (d) Applications and examples

Section 2. Elementary methods

- (a) Smooth variations
- (b) Calculation of areas of irregular shapes
- (c) Calculation of speeds along a curve

Section 3. Limit

- (a) Concepts of limits
- (b) Calculation of limits
- (c) Continuous functions

Examples and Applications

<u>Chapter 2.</u> <u>Derivatives and Differentiations</u>

Section 1. Concepts

- (a) Examples
- (b) Definition of differentiation
- (c) The rate of change--examples
- (d) Second derivative

Section 2. Geometric meanings of derivatives

- (a) Geometric meanings
- (b) Concave and convex
- (c) Approximations by triangles

Section 3. Calculation of derivatives

- (a) Some functions and their derivatives
  - (1) Constant function
  - (2) Exponential function
  - (3) Trigonometric function
  - (4) Logarithmic function
- (b) Calculations
  - (1) Derivatives of sum or difference of functions
  - (2) Derivatives of products of functions
  - (3) Derivatives of quotients of functions
- (c) Differentials

Section 4. Composition of functions

- (a) Concepts
- (b) Derivative of composition of functions
- (c) Examples

Section 5. Exponential functions and inverse trigonometric functions
- (a) Exponential functions and their derivatives
- (b) Inverse trigonometric functions
- (c) Summary
- (d) Implicit functions

Section 6. Maximum and minimum

Section 7. Approximation and application
- (a) Linear approximation and application
- (b) Calculation of approximations
- (c) Errors and their estimates

Examples and Applications

<u>Chapter 3</u>. <u>Integration</u>

Section 1. Concepts of definite integration
- (a) Examples
- (b) Definitions
- (c) Geometric meanings

Section 2. Relations between integration and differentiation
- (a) Some basic formulas
- (b) Properties of definite integration
- (c) Relation between integration and differentiation
- (d) Examples

Section 3. Concepts of indefinite integration
- (a) Concepts
- (b) Elementary formula and calculations
- (c) Integration table
- (d) Translation of functions
- (e) Integration by parts

Section 4. Applications
- (a) Calculation of length of curves
- (b) Pressure of liquid
- (c) Measure of electricity
- (d) Average value

Section 5. Approximation
- (a) Counting small squares
- (b) Trapezoid method

Applications

This book is somewhat similar to "Calculus I" in the United States. It introduces differentiation and integration with heavy calculations. It defines "limit" intuitively without any rigor. That makes it easy to read but it would not provide a solid background for the reader to do theoretic research.

Elementary Geometry (Group on Elementary Mathematics, Peking, 1975; a reference book intended for older students, people from the factory, and teachers in the middle school)

Chapter 1. Foundations

Section 1. Lines, angles, and circles
Section 2. Crossover and parallel
Section 3. Examples

Chapter 2. Triangles

Section 1. Sum of angles in a triangle
Section 2. Isomorphic triangles
Section 3. Isosceles triangles
Section 4. Right triangles
Section 5. Parallelogram and trapezoid

Chapter 3. Homeomorphic triangles

Section 1. Segments with constant ratio
Section 2. Homeomorphic triangles
Section 3. Applications

Chapter 4. Relations between angles and sides of a triangle

Section 1. Trigonometric functions
Section 2. Find angles and lengths in a right triangle

Chapter 5. Circles

Section 1. Elementary properties
Section 2. Divide a circle into arcs of equal lengths
Section 3. Tangent
- (a) Tangent lines of a circle
  - (1) Properties of tangent lines
  - (2) How to draw tangent lines
  - (3) Lines and a circle
- (b) Tangent of two circles
  - (1) Relation between two circles
  - (2) Arcs and circles

Section 4. Radians

Chapter 6. Plane and lines in 3-dimensional (Euclidean) spaces

Section 1. Fundamentary properties
Section 2. Parallel
Section 3. Intersection of a line and a plane
Section 4. Intersection of planes

Summary

In this book, some basic concepts in geometry are introduced. It gives some axioms without distinguishing them from theorems. It states that the Pythagorean theorem was proved in 1000 B.C., written in an ancient book in Chou dynasty. (Pythagoras lived in 600 B.C.) It emphasizes applications such as the measurements of land, the design of tools, etc.

Elementary Probability (Applied Mathematics Group, Department of Mathematics, Shanghai Normal College, 1975)

1. Random Phenomena Can Be Studied
   - 1.1 Weather forecast as a random phenomenon
   - 1.2 Characteristics of probabilistic methods
   - 1.3 Some typical examples of random phenomena
2. Random Events and Their Probabilities
   - 2.1 Random events
   - 2.2 The probability of the occurrence of an event
   - 2.3 The interdependences among events
3. The Classical Model
   - 3.1 The classical model
   - 3.2 Combinations and arrangements
   - 3.3 Examples for the computations of probabilities under the classical model
   - 3.4 The hypergeometric distribution and quality inspection
4. Random Variables
   - 4.1 What is a random variable?
   - 4.2 Discrete random variables
   - 4.3 Distributions for discrete random variables
   - 4.4 Expectations for random variables
5. Statistical Independence and Conditional Probabilities
   - 5.1 Condition probabilities and independent events
   - 5.2 The general multiplication law for probabilities
   - 5.3 Examples for the applications of independent probabilities
   - 5.4 A formula for mutually exclusive events and the Bayes rule
6. The Binomial Distribution and the Poisson Distribution
   - 6.1 *n* independent trials and the binomial distribution
   - 6.2 The Poisson distribution
   - 6.3 The inspection of the quality of drinking water
   - 6.4 The inference to the population from a sample
7. Variances of Random Variables
   - 7.1 Definition of variance
   - 7.2 Laws of large numbers
   - 7.3 Unbiased estimations for expectations and for variances
8. The Normal Distribution
   - 8.1 The frequency histogram and continuous random variables
   - 8.2 Properties of normal distribution
   - 8.3 Examples for the applications of normal distributions

The content of this book is not very different from the first 10 chapters of Feller's *Probability Theory*, except that it is not as in-depth. Probability is introduced from the frequentist's viewpoint and the general treatment of the subject is toward its application in statistics. Examples are given not only to illustrate a theoretical point, but more to demonstrate methods for real problems. Therefore, the urn model and the occupancy model so popular in Western texts are not mentioned once in this book.

Optimal Selection Algorithms (Optimal Seeking Group, Operations Research Department, Institute for Mathematics, Peking, 1975)

5.2 Complete bilevel experiments
5.3 Fractional bilevel experiments

6. Examples of Applications and Comments

6.1 How to set objectives
6.2 How to analyze the factors
6.3 How to define the operative regions of the factors
6.4 How to select algorithms
6.5 The scope of applications for optimal selection has been expanding

Appendix

1. Unimodal functions and their properties
2. Fraction algorithm is best
3. Simultaneous testing
4. Discussions on the multifactor case

References

This is a cookbook of methods for determining the optimal combination of factor levels. The menu stops at locating the optimal levels with very little regard to tests of significance. Examples are numerous and always downright practical.

Regression Analysis (Mathematical Statistics Group, Institute for Mathematics, Academy of Sciences, Peking, 1975)

1. What is Regression Analysis?

1.1 Two different types of relations between variables--function and correlation.
1.2 What kind of problems can regression analysis solve?

2. Solving the Regression Line

2.1 Scattergram and the regression line
2.2 Principles to determine the regression line
2.3 The computation formulas
2.4 Simplified computation for regression
2.5 Weighted regression

3. Correlation Coefficient and Test of Significance

3.1 The definition of correlation coefficient
3.2 The meaning of correlation coefficient
3.3 Significance test of correlation coefficient
3.4 A simplified method to obtain the correlation coefficient and the regression line

4. Test of Significance for Linear Regression

4.1 Analysis of variance for regression
4.2 Forecasting or controlling the Y value from the regression

The book deals with classical regression analysis. Since statistical distribution theory is not covered, the subject of hypothesis testing is barely touched except for a few cookbook formulas. Matrix terminology is not used even for multiple regression. One surprising fact is that the grand example thoroughly discussed in Section 9.3 is taken from Western texts, a rare exception to the overwhelming practice of using the case histories from local factories as examples.

Orthogonal Experimental Designs (Shanghai Normal College, 1975)

1. The Basic Ideas and Intuitive Analysis of Orthogonal Experimental Designs
    - 1.1 Introduction--Experiments with multifactor
    - 1.2 Using the orthogonal tables for experiments
    - 1.3 Interaction
    - 1.4 On degrees of freedom and the selection of orthogonal tables
    - 1.5 Designs and analyses for multifactor experiments when the factors have same level
2. Analysis of Variance
    - 2.1 The need for analysis of variance
    - 2.2 Analyses of variance for single-factor experiments
    - 2.3 Some mathematical concepts in analysis of variance
    - 2.4 Analyses of variance for multifactor experiments
    - 2.5 Analyses of variance for experiments with replication
3. Further Use of Orthogonal Tables in Experimental Designs
    - 3.1 Combining columns in an orthogonal table
    - 3.2 Condensing a high-level factor
    - 3.3 Methods for reinforcing an experiment
    - 3.4 Split-plot designs
4. Data Structure
    - 4.1 Structure for experimental data
    - 4.2 Estimating the effects in an orthogonal experimental design
    - 4.3 Missing data
5. Some Statistical Concepts in Experimental Designs
    - 5.1 Events and probability
    - 5.2 Normal distribution
    - 5.3 The expectation and variance of a normal distribution
    - 5.4 F-test
    - 5.5 A brief explanation for the statistical analysis in Chapter 3
6. Mixed Factorial Experiments
7. The Constructions for Orthogonal Tables
    - 7.1 Requirements for an orthogonal table
    - 7.2 Orthogonality
    - 7.3 The constructions for bilevel orthogonal tables
    - 7.4 The constructions for other orthogonal tables
8. Application Examples
    - 8.1 The use of orthogonal design in an antibiotic fermentation experiment

8.2 An experiment in seeking the best conditions to synthesize *PS* $(NH_2)_3$

Appendix

1. Orthogonal tables
2. Table of critical values for *F*-test

The subject of factorial and fractional factorial experimental designs is quite thoroughly covered, with the tone setting at practical level and not wandering too much into mathematical generality. A statistical background for conducting *F*-tests is provided. There are many examples from industry and agriculture.

Differential Equations and their Numerical Solutions (University text-book, Shanghai; prepared by Futan University)

1. Simple ordinary differential equations.

2. Vibration of a mass point.

3. Linear differential equations of higher order with constant coefficients.

4. Systems of linear differential equations with constant coefficients.

5. Numerical solution of the initial value problem of (systems of) ordinary differential equations.

6. Boundary value problem of ordinary differential equations.

7. Transverse vibrations of a beam.

8. Equation of wave motion.

9. Equation of heat conduction.

10. Harmonic equation.

11. Finite element method.